An Introduction to
Wavelet Theory
in **Finance**

A Wavelet Multiscale Approach

An Introduction to
Wavelet Theory
in Finance
A Wavelet Multiscale Approach

Francis In
Monash University, Australia

Sangbae Kim
Kyungpook National University, Korea

 **World Scientific**

NEW JERSEY · LONDON · SINGAPORE · BEIJING · SHANGHAI · HONG KONG · TAIPEI · CHENNAI

Published by

World Scientific Publishing Co. Pte. Ltd.

5 Toh Tuck Link, Singapore 596224

USA office: 27 Warren Street, Suite 401-402, Hackensack, NJ 07601

UK office: 57 Shelton Street, Covent Garden, London WC2H 9HE

Library of Congress Cataloging-in-Publication Data
In, Francis.
  An introduction to wavelet theory in finance : a wavelet multiscale
approach / by Francis In & Sangbae Kim.
    p. cm.
  Includes bibliographical references and index.
  ISBN 978-9814397834 -- ISBN 9814397830
  1. Finance--Mathematical models. 2. Wavelets (Mathematics) I. Kim, Sangbae, 1965–
II. Title.
  HG106.I5 2012
  332.01'5152433--dc23

                                                            2012030894

British Library Cataloguing-in-Publication Data
A catalogue record for this book is available from the British Library.

In-house Editor: Alisha Nguyen

Typeset by Stallion Press
Email: enquiries@stallionpress.com

Printed in Singapore.

Contents

Chapter 1

Methodology: Introduction to Wavelet Analysis

1.1. Introduction

The multiscale relationship is important in economics and finance because each investor has a different investment horizon. Consider the large number of investors who participate in the stock market and make decisions over *different time scales*. Stock market participants are a diverse group that include intraday traders, hedging strategists, international portfolio managers, commercial banks, large multinational corporations, and national central banks. It is notable that these market participants operate on very different time scales. In fact, due to the different decision-making time scales among traders, the true dynamic structure of the relationship between variables will *vary* over different time scales associated with those different horizons. However, most previous studies focus on a two-scale analysis — short-run and long-run. The reason being for this is mainly a lack of empirical tools. Recently, wavelet analysis has attracted attention in the fields of economic and finance as a means of filling this gap.

Wavelet analysis is relatively new in economics and finance, although the literature on wavelets is growing rapidly. The studies, related to economics and finance, can be divided into four categories: general wavelet transform, stationary process (long memory), denoising, and variance/covariance analysis. The first category includes Davidson *et al.* (1998), Pan and Wang (1998), Ramsey and Lampart (1998a, 1998b), and Chew (2001). Another stream of research is related to the long memory process of time series. Researchers in this field include Ramsey *et al.* (1995) for self-similarity (long memory); Jensen (1999b), Tkacz (2001), and Whitcher and Jensen (2000) for wavelet OLS; Jensen (1999a, 2000) for the wavelet maximum likelihood method; and Jamdee and Los (2006), Kyaw *et al.* (2006) and Manimaran *et al.* (2005, 2006) for the wavelet Hurst exponent.

The third category of recent wavelet analysis is the use of the wavelet denoising (detrending) method, for example, Fleming *et al.* (2000) and Capobianco (2003). The final category involves the application of wavelet analysis to multiscale variance/covariance analysis. This stream of wavelet analysis is mostly based on Percival and Walden (2000) and Gençay *et al.* (2002). Applications of this method include Gençay *et al.* (2001), Gençay *et al.* (2003, 2005) and In and Kim (2006).

The purpose of this chapter is to introduce wavelet analysis and to focus on what features of wavelet analysis can be applied to financial analysis. To examine wavelet analysis, this chapter begins with Fourier analysis, and then the chapter moves to the main features of wavelet analysis largely focusing on the application of time series analysis.

1.2. Fourier Analysis and Spectral Analysis

Fourier analysis and spectral analysis are used in modern signal processing and business cycle theory. This section introduces and investigates the properties of Fourier analysis and spectral analysis.

1.2.1. *Fourier analysis*

In the history of mathematics, wavelet analysis has many different origins. One of them is Fourier analysis. The fundamental idea in Fourier analysis is that any deterministic function of frequency can be approximated by an infinite sum of trigonometric functions, called the Fourier representation.

Fourier's result states that any function $f \in L^2[-\pi, \pi]$[1] can be expressed as an infinite sum of dilated cosine and sine functions:

$$f(x) = \frac{1}{2}a_0 + \sum_{j=1}^{\infty}(a_j \cos(jx) + b_j \sin(jx)) \tag{1.1}$$

---

[1]L^2 is a space of all functions with a well-defined integral of the sequence of the modulus of the function.

where an appropriately computed set of coefficients $\{a_0, a_1, b_1, \ldots\}$ is a complex sequence. As can be seen in Equation (1.1), in Fourier transform, any signal can be expressed as a function of sines and cosines. The Fourier basis functions (sines and cosines) are very appealing when representing a time series that does not vary over time, i.e., a stationary time series (Gençay *et al.*, 2002:97).

Equation (1.1) has to be interpreted with a caution. The equality is only meant in the L^2 sense, i.e.:

$$\int_{-\pi}^{\pi} \left[f(x) - \left(\frac{1}{2}a_0 + \sum_{j=1}^{\infty} (a_j \cos(jx) + b_j \sin(jx)) \right) \right]^2 dx = 0 \qquad (1.2)$$

It is possible that f and its Fourier representation differ on a few points (this is the case at discontinuity points). The summation in Equation (1.2) is up to infinity, but a function can be well-approximated (in the L^2 sense) by a finite sum with upper summation limit index J:

$$H_J(x) = \frac{1}{2}a_0 + \sum_{j=1}^{J} (a_j \cos(jx) + b_j \sin(jx)) \qquad (1.3)$$

This Fourier series representation is highly useful in that any L^2 function can be expressed in terms of two basis functions: sines and cosines. This is because of the fact that the set of functions $\{\sin(j\cdot), \cos(j\cdot), j = 1, 2, \ldots\}$, together with the constant function, form a basis for the function space $L^2[-\pi, \pi]$ (Ogden, 1997).

The Fourier basis has three important properties. The first property is that it has an orthogonal basis. Orthogonality implies that the inner product of any two functions $f_1, f_2 \in L^2[a, b]$ is equal to zero, resulting from the sine and cosine functions.

The second property of Fourier transform is orthonormality, which means that the sequence of function f_js are pairwise orthogonal and $\|f_j\| = 1$ for all j. Defining $u_j(x) = \pi^{-1/2} \sin(jx)$ for $j = 1, 2, \ldots$ and $v_j(x) = \pi^{-1/2} \cos(jx)$ for $j = 1, 2, \ldots$ with the constant function $v_0(x) = 1/\sqrt{2\pi}$ on $x \in [-\pi, \pi]$ makes the set of functions $\{v_0, u_1, v_1, \ldots\}$ orthonormal (Ogden, 1997).

Finally, the Fourier basis is a completely orthonormal system (Ogden, 1997). It is said that a sequence of function $\{f_j\}$ is a complete orthonormal system if the f_js are pairwise orthogonal, $\|f_j\| = 1$ for each j, and the only function orthogonal to each f_j is the zero function. Thus, the set $\{h_0, g_j, h_j : j = 1, 2, \ldots\}$ is a complete orthonormal system for $L^2[-\pi, \pi]$.

From Equation (1.1), we observe that the Fourier transform consists of sine and cosine functions, which are periodic functions. Therefore, if a function, $f(x)$, is a non-periodic signal, the expression of this function as the summation of the periodic functions (sine and cosine) does not accurately capture the movement of the signal. One could artificially extend the signal to make it periodic. However, it would require additional continuity at the endpoints. To avoid this problem, Gabor (1946) introduces the windowed Fourier transform (WFT, also called the short time Fourier transform) to measure the frequency variation of a signal. The WFT can be used to give information about signals simultaneously in the time and frequency domains. A real and symmetric window $u(t) = u(-t)$ is translated by k and modulated by the frequency ξ (Mallat, 1999):

$$u_{k,\xi}(t) = e^{i\xi t}u(t - k) \tag{1.4}$$

It is normalized $\|u\| = 1$ so that $\|u_{k,\xi}\| = 1$ for any $(k, \xi) \in \Re$. The resulting WFT of $f \in L^2(\Re)$ is:

$$Sf(u, \xi) = \langle f, u_{k,\xi} \rangle = \int_{-\infty}^{+\infty} f(t)u(t - k)e^{-i\xi t}dt \tag{1.5}$$

where $Sf(k, \xi)$ is the WFT. Therefore, with the WFT, the input signal $f(x)$ is divided into several sections, and each section is analyzed for its frequency content separately. The effect of the window is to localize the signal in time. The WFT represents a sort of compromise between the time- and frequency-based views of a signal. It provides some information about both when and at what frequencies a signal event occurs. However, we can only obtain this information with limited precision, determined by the size of the window. While the WFT compromise between time and frequency information can be useful, the drawback is that once you choose a particular size for the time window, that window is the same for all frequencies. Another drawback of the WFT is that it will not be able to resolve events if they happen to appear within the width of the window (Gençay *et al.*, 2002:99).

Another extension of the Fourier transform is the Fast Fourier Transform (FFT). To approximate a signal using the Fourier transform requires application of a matrix, the order of which is the number of sample points n. Since multiplying an $n \times n$ matrix by a vector costs in the order of n^2 arithmetic operations, the computational burden increases enormously with the number of sample points. However, if the samples are uniformly spaced, then the Fourier matrix can be factored into a product of just a few sparse matrices, and the resulting factors can be applied to a vector in a total of order $n \log n$ arithmetic operations. This is the so-called FFT (Graps, 1995).

1.2.2. *Spectral analysis*[2]

As discussed earlier, Fourier analysis transforms time domain data into frequency domain data. This feature naturally leads researchers to look for ways of determining which are dominant frequencies in a time series. In our research we have adopted one method that has proved successful in much researches, namely, power spectral density. The power spectral density of a time series x_t, denoted in $S_x(w)$, can be defined as the Fourier transform of the autocorrelation function.

$$S_x(w) = \int_{-\infty}^{\infty} R_x(\tau)e^{-j2\pi w\tau}d\tau \quad \text{where } R_x(\tau) = E\{x_t x_{t+\tau}\} \quad (1.6)$$

For example, suppose that we compute the power spectra of a covariance stationary stochastic process when we know the stochastic process, which has generated the time series. Note that any covariance stationary stochastic process can be expressed as an infinite moving average representation or *Wold representation* as:

$$x_t = \psi(L)\varepsilon_t = \sum_{j=0}^{\infty} \psi_j L^j \varepsilon_t \quad (1.7)$$

where $\psi(L) = 1 + \psi_1 L + \psi_2 L^2 + \cdots$ is a polynomial in the lag operator, L, and $\psi_0 = 1$. The spectrum of the white noise process is $S_\varepsilon(w) = \sigma_\varepsilon^2/2\pi$, and Equation (1.7) shows that x_t is generated by filtering the white noise process where $\psi(L)$ are the filter weights. The spectrum of x_t is thus the spectrum of the white noise process multiplied by the effect of the filter (Pedersen, 1999). This is easily computed taking the following steps:

1. Formulating the model in terms of its moving average representation (1.7) using the lag operator.

---

[2]In relation to the business cycle, the literature includes Howrey (1968), Sargent and Sims (1977), Baxter and King (1999), and recently Sarlan (2001), while in finance, the literature includes examination of the stock market (Fischer and Palasvirta, 1990; Knif *et al.*, 1995; Lin *et al.*, 1996; Asimakopoulos *et al.*, 2000; Smith, 1999 and 2001), and investigation of the interest rates (Kirchgassner and Wolters, 1987; Hallett and Richter, 2001, 2002). For a more extensive survey of the application of spectral analysis to economics and finance, see Ramsey and Thomson (1999).

2. Substitute e^{-wt} for the lag operator L to get

$$x_t = \sum_{j=0}^{\infty} \psi_j L^j \varepsilon_t = (1 + \psi_1 e^{-iw} + \psi_2 e^{-2iw} + \psi_3 e^{-3iw} + \cdots)\varepsilon_t$$

This is the transfer function or the frequency response function of the filter (1.7).

3. Take the square of the absolute value of the frequency response function, called the power transfer function of the filter, denoted as:

$$H(w) = |1 + \psi_1 e^{-iw} + \psi_2 e^{-2iw} + \psi_3 e^{-3iw} + \cdots|$$

4. The power spectral density function is the power transfer function of the filter multiplied by the spectrum of the white noise process

$$S_x(w) = H(w) \cdot S_\varepsilon(w)$$

$$= \frac{1}{2\pi}|1 + \psi_1 e^{-iw} + \psi_2 e^{-2iw} + \psi_3 e^{-3iw} + \cdots|\sigma_\varepsilon^2$$

For example, consider the following ARMA(2,2) process:

$$x_t = \theta_1 x_{t-1} + \theta_2 x_{t-2} + \varepsilon_t + \phi_1 \varepsilon_t + \phi_2 \varepsilon_{t-2}$$

To compute the power spectrum density, we follow four steps mentioned above.

Step 1. Reformulate above equation in terms of its moving average representation using the lag operator.

$$x_t = \frac{1 + \phi_1 L + \phi_2 L^2}{1 - \theta_1 L - \theta_2 L}\varepsilon_t$$

Step 2. Substitute for the exponential function instead of the lag operator.

$$x_t = \frac{1 + \phi_1 e^{-iw} + \phi_2 e^{-2iw}}{1 - \theta_1 e^{-iw} - \theta_2 e^{-2iw}}\varepsilon_t$$

Step 3. The corresponding frequency response function can be defined as follow:

$$h(e^{-iw}) = \frac{1 + \phi_1 e^{-iw} + \phi_2 e^{-2iw}}{1 - \theta_1 e^{-iw} - \theta_2 e^{-2iw}}$$

and the power transfer function becomes:

$$H(w) = \left|\frac{1 + \phi_1 e^{-iw} + \phi_2 e^{-2iw}}{1 - \theta_1 e^{-iw} - \theta_2 e^{-2iw}}\right|^2$$

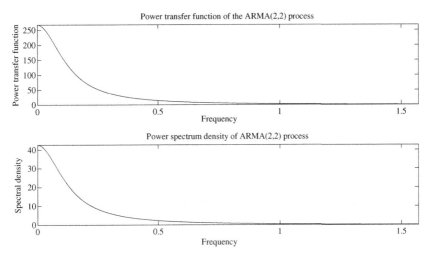

Fig. 1.1.   The power transfer function and the power spectral density function of ARMA (2, 2) process.
Note: This figure plots the power transfer function and the power spectral density function of ARMA(2, 2) process. The upper figure presents the power transfer function, calculated by $H(w)$ in Step 3. The bottom figure presents the power spectral density function calculated by Step 4.

Step 4. Finally, the power spectral density function can be defined as:

$$S_x(w) = \frac{1}{2\pi} \left| \frac{1 + \varphi_1 e^{-iw} + \varphi_2 e^{-2iw}}{1 - \theta_1 e^{-iw} - \theta_2 e^{-2iw}} \right|^2 \sigma_\varepsilon^2$$

The power spectral density function of this ARMA(2, 2) process is plotted in Fig. 1.1 with the power transfer function.

This spectral analysis can be applied to the multivariate case. It is known as cross spectral analysis, and it allows us to examine the multivariate case in the frequency domain. This presents an alternative method to investigate the lead-lag relationship and comovements between time series. The cross spectrum is a complex quantity and can be reformed in terms of two real quantities, the cospectrum, $co_{ij}(w)$, and the quadrature spectrum, $qu_{ij}(w)$. Using Euler Relations and DeMoire's Theorem, the coherence can be decomposed into two terms as follows.

$$S_{ij} = co_{ij}(w) + iqu_{ij}(w) \tag{1.8}$$

The cospectrum can be expressed by:

$$co_{ij}(w) = \frac{1}{2\pi} \sum_{s=-\infty}^{\infty} R_{ij}(s)\cos(ws) \tag{1.9}$$

where R_{ij} is defined as $E[(i_t - Ei_t)(j_t - Ej_t)]$. The cospectrum between two series i and j at frequency w can be interpreted as the covariance between two series i and j that is attributable to cycles with frequency w. The cospectrum can have either positive or negative values, since the autocovariances can be both positive and negative.

The quadrature spectrum is rewritten as follows:

$$qu_{ij}(w) = \frac{1}{2\pi} \sum_{s=-\infty}^{\infty} R_{ij}(s)\sin(ws) \tag{1.10}$$

The quadrature spectrum from time series i to time series j at frequency w is proportional to the portion of the covariance between two time series i and j due to cycles of frequency w. From this quadrature spectrum, we can observe which series has more out-of-phase cycles, because time series i may respond to an economic recession later than time series j.

Next, we need to show how to derive the gain in spectral analysis. The gain has a feature which shows how a change of the regression coefficients in the time domain can affect the cross spectrum.

$$|G(w)| = \sqrt{(co_{ij}(w))^2 + (qu_{ij}(w))^2} \tag{1.11}$$

The function $|G(w)|$ is called the gain. The gain is equivalent to the regression coefficient for each frequency w. In other words, it measures the amplification of the frequency components of the j-process to obtain the corresponding components of the i-process.

The estimated coherence spectrum between two series for various frequencies is given by:

$$Coh(w) = \frac{|S_{ij}(w)|^2}{S_{ii}(w)S_{jj}(w)} \tag{1.12}$$

where S_{ii} and $S_{jj}(i \neq j)$ are the autospectrum estimated from:

$$S_{ii}(w) = \frac{1}{2\pi} \sum_{-(N-1)}^{N-1} \lambda_N(s)R_{ii}(s)e^{-iws} \tag{1.13}$$

where R_{ii} is defined as $E[(i_t - Ei_t)^2]$. The coherence is a real-valued function, which has a value between 0 and 1. The coherence between two time series measures the degree of which series are jointly influenced by cycles at frequency w. In other words, as can be seen Equation (1.12), coherence is the ratio of the squared cross spectrum to the product of two autospectrums, analogue to the squared coefficient of correlation. We can use the coherence between two or more time series to measure the extent to which multiple time series move together over the business cycle.

The lead-lag relationship between two time series can be captured by the phase. The phase, defined as $\varphi(w)$, can be expressed as a ratio between the cospectrum and quadrature spectrum.

$$\varphi(w) = \tan^{-1}\left(\frac{co_{ij}(w)}{qu_{ij}(w)}\right) \qquad (1.14)$$

In addition, the phase gives the lead of one series over another series at frequency w. The phase graph gives information about the lead-lag relationship between two time series. If the phase is a straight line over some frequency band, the slope is equal to the time lag and thus tells which series is leading and by how many periods (Pederson, 1999).

This reveals the lead and lag relationship between two variables at different frequencies. In other words, a positive phase slope indicates that the input variable leads the output variables, while a negative phase slope indicates that the input variable lags.

1.2.3. *Comparison between Fourier transform and wavelet transform*

While Fourier analysis is one of the origins of wavelet analysis, the two methods have some points of similarity and some important points of difference. The first similarity is reversibility. Both transforms are reversible functions. That is, they allow going back and forward between the raw and transformed signals. Another similarity is that the basis functions are localized in frequency, making mathematical tools such as power spectra (how much power is contained in a frequency interval) and scalograms useful at picking out frequencies and calculating power distributions (Graps, 1995).

Even though they possess some similarities, the two transforms are different from each other. Through examining the difference between the two transforms, we can easily see why we need wavelet analysis instead of Fourier analysis. Wavelet analysis has three distinctive advantages over Fourier analysis. The first advantage is that wavelet analysis has the ability to decompose the data into several time scales instead of the frequency domain. This advantage allows us to examine the behavior of a signal over various time scales. The second advantage of wavelet transforms is that the windows vary. In order to isolate signal discontinuities, one would like to have some very short basis functions. At the same time, in order to obtain detailed frequency analysis, one would like to have some very long basis functions. In fact, wavelet transforms allow us to do both. The final advantage of wavelet transforms is their ability to handle the non-stationary data. Restricting to stationary time series would not be very promising and appealing since most interesting time series exhibit quite complicated patterns over time (trends, abrupt regime changes, bursts of variability, etc.).

To compare Fourier transform and wavelet transform, we decompose the sine signal $\{s_t = \sin(\pi \cdot t/1.5)\}$ using Fourier transform and continuous wavelet transform. To calculate the wavelet coefficients, the Haar wavelet filter has been adopted. More properties of the Haar wavelet filter will be discussed later. Fig. 1.2 plots the original sine signal, Fourier coefficients, and wavelet coefficients.

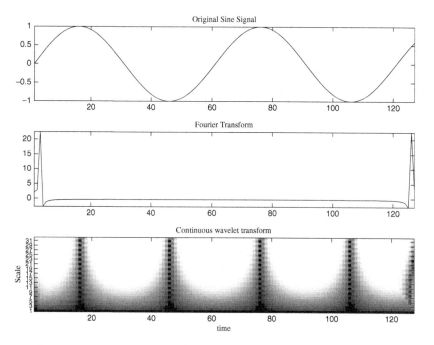

Fig. 1.2.   Comparison between Fourier transform and wavelet transform.
Note: This figure illustrates the original sine signal, the Fourier transform, and the wavelet transform. The second figure plots the Fourier transform of the original signal. Clearly it indicates the original signal has a single frequency. The bottom figure indicates the continuous wavelet transform. This figure has been constructed using Haar wavelet filter, which will be discussed later in this chapter.

As can be seen in Fig. 1.2, a plot of the Fourier coefficients of this signal shows nothing particularly interesting: a flat spectrum with two peaks represents a single frequency. More specifically, the Fourier transform picks up the low-frequency oscillation and lacks strong evidence of the discontinuity. In other words, Fourier analysis only shows the global movements, not local movements. However, by giving up some frequency resolution, the wavelet transform has the ability to capture events that are local in time. This makes the wavelet transform an ideal tool for studying nonstationary or transient time series. In contrast to the Fourier transform, as shown in Fig. 1.2, the wavelet transform clearly identifies the abrupt change in the function and the low-frequency sinusoid.

1.3. Wavelet Analysis

As a means of understanding the fundamentals of wavelet analysis, Daubechies (1992) provides an extensive look at the mathematical properties of wavelets. Chui (1992), and Strang and Nguyen (1996) are good introductions to wavelets. The text by Gençay *et al.* (2002) gives a good discussion on how wavelets can be applied in economics and finance. Ramsey (1999, 2002) and Schleicher (2002) also give some additional insights on how wavelet analysis can be adopted in economics and finance. In this section, we examine the properties of the continuous wavelet transform, and two discrete wavelet transforms (discrete wavelet transform and maximal overlap discrete wavelet transform).

1.3.1. *Continuous wavelet transform*

The continuous wavelet transform (CWT) is defined as the integral over all time of the signal multiplied by scaled, shifted versions of the wavelet function ψ (*scale, position, time*):

$$C(scale, position) = \int_{-\infty}^{\infty} x_t \psi(scale, position, t)dt \qquad (1.15)$$

The results of the CWT are many wavelet coefficients C, which are a function of scale and position. The scale and position can take on any values compatible with the region of the time series x_t. Multiplying each coefficient by the appropriately scaled (dilated) and shifted wavelet yields the constituent wavelets of the original signal. If the signal is a function of a continuous variable and a transform that is a function of two continuous variables is desired, the continuous wavelet transform (CWT) can be defined by (Burrus *et al.*, 1998):

$$F(a, b) = \int x_t \psi \left(\frac{t - a}{b} \right) dt \tag{1.16}$$

with an inverse transform of:

$$x_t = \iint F(a, b) \psi \left(\frac{t - a}{b} \right) da\, db \tag{1.17}$$

where $\psi(t)$ is the basic wavelet and $a, b \in \mathrm{R}$ are real continuous variables.

To capture the high and low frequencies of the signal, the wavelet transform utilizes a basic function (mother wavelet) that is stretched (scaled) and shifted.

Scale of wavelets

Scaling a wavelet simply means stretching (or compressing) it. To go beyond colloquial descriptions such as "stretching", we introduce a scale factor, b, so that $\psi_b(t) = \psi(t/b)$. The smaller the scale factor, the more "compressed" the wavelet (see Fig. 1.3). It is natural to think about a correspondence between wavelet scales and frequency. A fine and small scale b generates a compressed wavelet. In turn, this compressed wavelet makes the details change rapidly. In consequence, a fine and small scale b can capture a high frequency oscillation. In contrast, a coarse and large scale b can capture low frequency movements.

Shifting of wavelets

Shifting a wavelet simply means advancing or delaying it. Mathematically, delaying a function $\psi(t)$ by a is represented by $\psi(t - a)$.

Using these two properties, the wavelet transform intelligently adapts itself to capture features across a wide range of frequencies and thus has the ability to capture events that are local in time.

What conditions must wavelets satisfy?

A wavelet $\psi(t)$ is a simple function of time t that obeys some rules (admissibility, orthogonality, vanishing moments). First, the admissibility condition is:

$$C_\psi = \int_0^\infty \frac{|H(w)|}{w} dw < \infty$$

where $H(w)$ is the Fourier transform, a function of frequency w, of $\psi(t)$ in the CWT. This condition is only useful in theoretical analysis, and as in the Fourier transform, there is a necessary condition to satisfy the

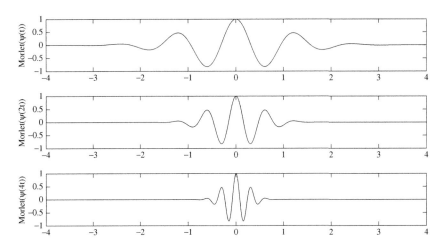

Fig. 1.3.   Morlet wavelet with different scales.
Note: This figure illustrates the effect of changing the scale *b*. From this figure, it is observed that the smaller value of *b* generates the more compressed wavelet filter.

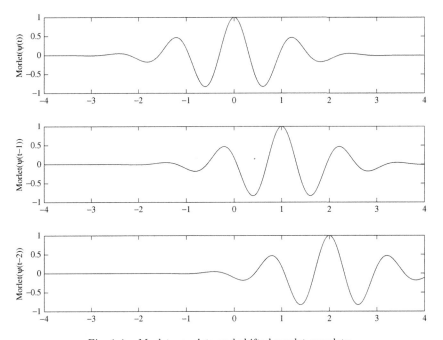

Fig. 1.4.   Morlet wavelets and shifted morlet wavelets.
Note: This figure presents the effect of changing *a* in Equation (2.17).

Dialect condition (i.e., continuous or only limited discontinuous points in the integration span). This condition ensures that $H(w)$ goes to zero as $w \to 0$ (Grossman and Morlet, 1984; Mallat, 1999).

In other words, to guarantee that $C_\psi < \infty$, we must impose the condition, $H(0) = 0$. This condition leads us to the first condition of a wavelet function.

$$\int_{-\infty}^{\infty} \psi(t)dt = 0 \tag{1.18}$$

If the energy of a function is defined as the squared function integrated over its domain, the second condition is that the wavelet function has unit energy.

$$\int_{-\infty}^{\infty} |\psi(t)|^2 dt = 1 \tag{1.19}$$

The second condition is orthogonality. As discussed in Fourier analysis, the wavelet also has an orthogonal property. Orthogonality means that the shifted functions in the same scale are orthogonal and also that the functions at different scales are orthogonal. In the implementation of a wavelet system by filter banks, orthogonality means that if the inverse of the analysis filter banks is exactly the transpose of itself, this wavelet is orthogonal. In this case, only one wavelet mother function is necessary for both analysis and synthesis. It does not have linear phase. If the wavelet is biorthogonal, the inverse of analysis filter bank is not necessarily the transpose of itself. In other words, there would be two mother functions for analysis and synthesis respectively. Some wavelet families can be both biorthogonal and orthogonal.

The final condition is related to vanishing moments. What is the relevance of vanishing moments? More vanishing moments means that the scale function is smoother. The number of vanishing moments comes from the defining wavelet equation.

1.3.2. *Discrete wavelet transform*

In time series analysis, the data has a finite length of duration. Therefore, only a finite range of scales and shifts are meaningful. In this section, we study the discrete wavelet transform (DWT).

To describe the idea of multiresolution, it is better to start from the properties of scale function (father wavelet). A two-dimensional family of functions is generated from the basic scaling function by scaling and translation as follows:

$$\phi_{j,k}(t) = 2^{-\frac{j}{2}}\phi(2^{-j}t - k) = 2^{-j/2}\phi\left(\frac{t - 2^j k}{2^j}\right) \qquad (1.20)$$

where 2^j is a sequence of scales. The term $2^{-j/2}$ maintains the norm of the basis functions $\phi(t)$ at 1. In this form, the wavelets are centered at $2^j k$ with scale 2^j. $2^j k$ is called the translation (shift) parameter. The change in j and k changes the support of the basis functions. 2^j is called the scale factor used for frequency partitioning. When j becomes larger, the scale factor 2^j becomes larger, and the function $\phi_{j,k}(t)$ becomes shorter and more spread out, and conversely when j gets smaller. Therefore, 2^j is a measure of the scale of the functions $\phi_{j,k}(t)$. The translation parameter $2^j k$ is matched to the scale parameter 2^j in the sense that as the function $\phi_{j,k}(t)$ gets wider, its translation step is correspondingly larger. This scaling function spans a space vector over k.

$$S_j = Span\{\phi_k(2^j t)\} \qquad (1.21)$$

In order to describe multiresolution analysis (MRA) more specifically, the basic requirement of MRA can be formulated as follows:

$$\cdots \subset S_{-2} \subset S_{-1} \subset S_0 \subset S_1 \subset S_2 \subset \cdots \subset L^2 \qquad (1.22)$$

with $S_{-\infty} = \{0\}$ and $S_\infty = L^2$.

Each subspace S_j encodes the information of the signal at resolution level j, which can be represented by scale functions (Lee and Hong, 2001). This relationship, plotted in Fig. 1.5, indicates that the space, which contains high resolution, also contains those of lower resolution.

From Fig. 1.5, we can find a relationship between two adjacent scaling functions such that if $\phi(t)$ is in V_0, it is also in V_1. This implies that $\phi(t)$ can be expressed as a weighted sum of shifted $\phi(2t)$. Therefore, MRA involves successively projecting all time series x_t to be studied into each of the approximation subspaces S_j.

$$\phi(t) = \sum_k g(k)\sqrt{2}\phi(2t - k), \quad k \in Z \qquad (1.23)$$

where the coefficients $g(k)$ are a sequence of real (complex) numbers called the scaling function coefficients (low-pass filter) and $\sqrt{2}$ maintains the norm of the scaling function with the scale of two. This equation is called the

$$S_0 \subset S_1 \subset S_2 \subset S_3$$

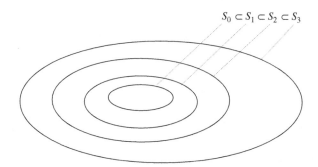

Fig. 1.5.   Nested vector spaces spanned by the scaling functions.
Note: This figure illustrates the vector spaces spanned by the scaling functions. From this figure, it is observed that the higher scaling function nests the lower scaling functions.

refine equation, the MRA equation, or the dilation equation as it describes different interpretations or points of view (Burrus *et al.*, 1998).

To this point, we have discussed some properties of scaling functions. These properties play an important role in describing the properties of the wavelet functions [mother wavelet, $\psi(t)$]. The important features of time series can be captured better by defining a slightly different set of functions $\psi(t)$ that span the differences between two adjacent spaces, spanned by the various scales of the scaling functions. The relationship between vector spaces of the scaling functions and those of wavelet functions are plotted in Fig. 1.6, based on the orthogonality condition of the scaling and wavelet functions (see Section 1.3.1). This condition gives several advantages. Orthogonal basis functions allow simple calculation of expansion coefficients and have a Parseval's theorem that allows a partitioning of the signal energy in the wavelet transform domain (Burrus *et al.*, 1998:4).

Combining this orthogonality with Fig. 1.6, we can describe L^2 as follows:

$$L^2 = S_0 \oplus D_1 \oplus D_2 \oplus D_3 \oplus \cdots \qquad (1.24)$$

where $\oplus$ denotes the orthogonal sum.

In Equation (1.24), we can describe the relation of S_0 to the wavelet spaces as follows:

$$S_0 = D_{-\infty} \oplus \cdots \oplus D_{-1} \qquad (1.25)$$

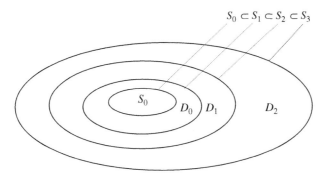

Fig. 1.6. Scaling function and wavelet vector spaces.
Note: This figure illustrates the relationship between scaling functions and wavelet functions.

This relationship shows that the key idea of MRA consists in studying a signal by examining its increasingly coarser approximations as more and more details are cancelled from the data (Abry *et al.*, 1998).

Analogous to this relationship, the wavelets reside in the space spanned by the next narrower scaling function: i.e., $D_0 \subset S_1$. This leads us to express the wavelets as a weighted sum of shifted scaling function $\phi(2t)$, which is defined in Equation (1.23), for some coefficients $h(k)$ (high-pass filter):

$$\psi(t) = \sum_k h(k)\sqrt{2}\phi(2t - k) \tag{1.26}$$

The function generated by Equation (1.26) gives the mother wavelet $\psi(t)$, which has the following form[3]:

$$\psi_{j,k}(t) = 2^{-\frac{j}{2}}\psi(2^{-j}t - k) = 2^{-j/2}\psi\left(\frac{t - 2^j k}{2^j}\right) \tag{1.27}$$

According to Equation (1.24), any time series $x_t \in L^2$ could be written as a series expansion in terms of the scaling function and wavelets.

$$f(t) = \sum_{k=-\infty}^{\infty} s(k)\phi_k(t) + \sum_{j=0}^{\infty}\sum_{k=-\infty}^{\infty} d(j,k)\psi_{j,k}(t) \tag{1.28}$$

---

[3]Intuitively, a small j or a low resolution level can capture smooth components of the signal, while a large j or a high resolution level can capture variable components of the signal (Lee and Hong, 2001).

In this expression, the first expansion gives a function that is a low resolution or a coarse approximation of x_t. For each increasing index j in the second summation, a higher or finer resolution function is added. The coefficients in this wavelet expansion are called the discrete wavelet transform (DWT[4]) of the signal x_t.

In another expression, the signal can be expressed as the sum of a finite set of high frequency parts and a residual low frequency part. The orthogonal wavelet series approximation up to scale J to a time series x_t is given by:

$$x_t \approx \sum_k s_{J,k}\phi_{J,k}(t) + \sum_k d_{J,k}\psi_{J,k}(t) + \sum_k d_{J-1,k}\psi_{J-1,k}(t)$$

$$+ \cdots + \sum_k d_{1,k}\psi_{1,k}(t) \tag{1.29}$$

with

$$s_{J,k} = \int \phi_{J,k}(t)x_t dt \quad \text{and} \quad d_{j,k} = \int \psi_{j,k}(t)x_t dt \quad \text{where } j = 1, 2, \ldots, J.$$

where J is the number of scales, and k ranges from 1 to the number of coefficients in the specified component. The coefficients $s_{J,k}, d_{J,k}, \ldots, d_{1,k}$ are the wavelet transform coefficients. J is the maximum integer such that 2^j is less than the number of data points. Their magnitude gives a measure of the contribution of the corresponding wavelet function to the approximation sum and wavelet series coefficients approximately specify the location of the corresponding wavelet function. More specifically, $s_{J,k}$ represents the smooth coefficients that capture the trend, while the detail coefficients $d_{J,k}, \ldots, d_{1,k}$, which can capture the higher frequency oscillations, represent increasing finer scale deviations from the smooth trend.

Given these coefficients, the wavelet series approximation of the original time series x_t is given by the sum of the smooth signal $S_{J,k}$, and the detail signals

$$D_{J,k}, D_{J-1,k}, \ldots, D_{1,k}:$$
$$x_t = S_{J,k} + D_{J,k} + D_{J-1,k} + \cdots + D_{1,k} \tag{1.30}$$

---

[4]Similar to the continuous wavelet transform (CWT), the DWT is a two dimensional orthogonal decomposition of a time series that is well suited, and is in fact designed, to detect abrupt changes and fleeting phenomena. The important characteristic of the DWT is that its basis functions have compact support. Thus, they are able to pick up unique phenomena in the data (Goffe, 1994).

where

$$S_{J,k} = \sum_k s_{J,k} \phi_{J,k}(t),$$

$$D_{J,k} = \sum_k d_{J,k} \psi_{J,k}(t)$$

and

$$D_{j,k} = \sum_k d_{j,k} \psi_{j,k}(t), \quad j = 1, 2, \ldots, J - 1$$

The original signal components $S_{J,k}, D_{J,k}, D_{J-1,k}, \ldots, D_{1,k}$ are listed in the order of increasingly finer scale components. Signal variations on high scales are acquired using wavelets with large supports.

The DWT maps the vector $f = (f_1, f_2, \ldots, f_n)'$ to a vector of n wavelet coefficients $w = (w_1, w_2, \ldots, w_n)'$. The vector w contains the coefficients $s_{J,k}, d_{J,k}, \ldots, d_{1,k}, j = 1, 2, \ldots, J$ of the wavelet series approximation, Equation (1.29). The DWT is mathematically equivalent to multiplication by an orthogonal matrix W:

$$w = Wf \tag{1.31}$$

where the coefficients are ordered from coarse scales to fine scales in the vector w. In the case where n is divisible by 2^J:

$$w = \begin{pmatrix} s_J \\ d_J \\ d_{J-1} \\ \vdots \\ d_1 \end{pmatrix} \tag{1.32}$$

where

$$s_J = (s_{J,1}, s_{J,2}, \ldots, s_{J,n/2^J})'$$
$$d_J = (d_{J,1}, d_{J,2}, \ldots, d_{J,n/2^J})'$$
$$d_{J-1} = (d_{J-1,1}, d_{J-1,2}, \ldots, d_{J-1,n/2^J})'$$
$$\vdots \qquad\qquad \vdots$$
$$d_1 = (d_{1,1}, d_{1,2}, \ldots, d_{1,n/2^J})'$$

Each set of coefficients $s_J, d_J, d_{J-1}, \ldots, d_1$ is called a crystal. The term crystal is used because the wavelet coefficients in a crystal correspond to a set of translated wavelet functions arranged on a regular lattice.

An alternative way to think about the wavelet is to consider low- and high-pass filters, denoted in Equations (1.23) and (1.26). The natural question is how to derive these filters so that they can be applied in wavelet analysis. The low- and high-pass filters can be obtained from the father and mother wavelets using the following relationships:

$$g(k) = \frac{1}{\sqrt{2}} \int \phi(t)\phi(2t - k)dt \qquad (1.33)$$

$$h(k) = \frac{1}{\sqrt{2}} \int \psi(t)\psi(2t - k)dt \qquad (1.34)$$

or

$$h(k) = (-1)^k g(k) \qquad (1.35)$$

The relationship between filter banks and wavelets is extensively discussed in Strang and Nguyen (1996) and Percival and Walden (2000). The analysis indicates that one can approach the analysis of the properties of wavelets either through wavelets or through the properties of the filter banks (Ramsey, 2002). However, the introduction of filter banks reveals clearly the difficulty of handling boundary conditions, which will be discussed in the next section more extensively.

It is important to examine the properties of the wavelet filter, in a similar manner to the continuous case. The DWT has counterpart properties to the case presented in Equations (1.18) and (1.19), which show integration to zero and unit energy. Let $h_l = (h_0, h_1, h_2, \ldots, h_{J-1})$ be a finite length discrete wavelet filter. This wavelet filter holds the same properties as the continuous wavelet function in as much as it sums to zero and has unit energy.

$$\sum_{l=0}^{J-1} h_l = 0 \quad \text{and} \quad \sum_{l=0}^{J-1} h_l^2 = 1$$

In addition to these properties, since the wavelet filters are orthogonal to its shifts, the following property also has to hold for all wavelet filters.

$$\sum_{l=0}^{J-1} h_l h_{l+2n} = 0, \quad \text{for all non-zero integers } n. \qquad (1.36)$$

This implies that to construct the orthonormal matrix that defines the DWT, wavelet coefficients cannot interact with one another.

Up to now, we have studied the DWT. To get an idea about how the DWT is applied in the signal (we use the stock index as an example). We apply the DWT to the daily S&P500 stock prices from June 29, 2000 to December 29, 2000. The return series are computed via the first difference of log-transformed prices — that is, $r_t = \log(P_t) - \log(P_{t-1})$. This series is plotted in the upper row of Fig. 1.7. There is an obvious increase in variance in the returns toward the latter half of the series. The length of the returns series is $N = 124$, which is divisible by $2^5 = 32$, and therefore, we may decompose our returns series up $J = 5$.

The wavelet coefficient vectors $d_1, \ldots, d_5$ using the Haar wavelet are shown in the left hand side of Fig. 1.7. The first scale of wavelet coefficients d_1 is filtering out the high-frequency fluctuations by essentially looking

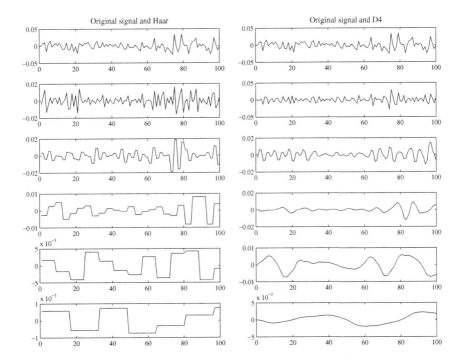

Fig. 1.7.   MRA using Haar and Daubechies 4 wavelets.
Note: This figure plots MRA using two different wavelet filters [Haar and D(4)] from June 29, 2000 to December 29, 2000 using daily frequency. The left hand side of this figure is constructed using the Haar wavelet filter, while the right-hand of this figure is constructed by the Daubechies wavelet filter with length 4. From this figure, we can observe that the D(4) wavelet filter generates more smooth wavelet coefficients than the Haar wavelet filter.

at adjacent differences in the data. There is a large group of rapidly fluctuating returns between observations from 60 to 100. A small increase in the magnitude is also observed between 60 and 100, but smaller than the unit scale coefficients. This vector of wavelet coefficients is associated with changes of λ_1, equivalent to 2–4 days. Since the S&P500 return series exhibits low-frequency oscillations, the higher (low-frequency) vectors of wavelet coefficients d_4 and d_5 indicate large variations from zero. Interestingly, as we noted above, the Haar wavelet is called as a step functions. As the wavelet time scale increases, the decomposed higher wavelet coefficients have a step-shape.

The same decomposition was performed using the Daubechies extremal phase wavelet filter of length 4 [D(4)] and provided in Fig. 1.7. The interpretations for each of the wavelet coefficient vectors are the same as in case of the Haar wavelet filter. The wavelet coefficients will be different given that the length of the filters is now four versus two, and should isolate feature in a specific frequency interval better since the D(4) is a better approximation over the Haar wavelet. Compared with the Haar wavelet coefficients, the wavelet coefficients of D(4) are smoother than those of the Haar wavelet, as the wavelet time scale increases.

In practice, the DWT is implemented through a pyramid algorithm (Mallat, 1989), which starts with a time series x_t. The first step of the pyramid algorithm is to use the wavelet filter and scaling filter to decompose the time series against various wavelet scales. During this procedure, the time series are down-sampled by two. Suppose that the original signal x_t consists of $N(= 500)$ samples of data. Then the approximation and the detail signals will have 1000 samples of data, for a total of $2N$ samples. To improve this sampling efficiency, we perform down-sampling. This simply means throwing away every second data point, down-sampled by 2, to produce the length $N/2^j$ wavelet coefficients vector d_j. In down-sampling, there is obviously the possibility of losing information, since half of the data is discarded. The effect in the frequency domain (Fourier transform) is called aliasing, which states that the result of this loss of information is mixing up of frequency components (Burrus *et al.*, 1998).

Fig. 1.8 gives a flow diagram for the first step of the pyramid algorithm, i.e., down-sampling. The symbol ↓ 2 implies that every second value of the time series vector is removed. More precisely, Fig. 1.8 illustrates the decomposition of x_t into the unit wavelet coefficients d_1 and the unit scale scaling coefficients s_1. The time series x_t is filtered using the wavelet filter and scaling filter and down-sampled by 2. Therefore, the N length vector of observations has been high- and low-pass filtered to obtain $N/2$ coefficients.

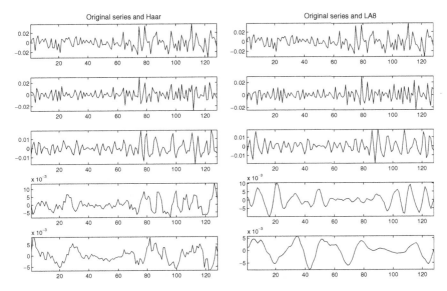

Fig. 1.8. Flow diagram illustrating the down-sampling.
Note: This figure illustrates how a time series is down-sampled using the high- and low-pass filters and shows how to obtain the wavelet and scaling coefficients using a pyramid algorithm.

The second step of the pyramid algorithm is to treat the scaling coefficients series $\{s_1\}$ as our original time series, and repeat the filter and down-sampling procedure using wavelet and scaling filters. In other words, this decomposition process can be iterated, with successive approximations being decomposed in turn, so that the time series is broken down into many lower resolution components. For example, suppose that one wants to decompose the time series x_t into a third level. After decomposing and down-sampling the original time series in the first level, shown in Fig. 1.8, the scaling coefficients are decomposed and down-sampled as we did in the original time series. Once we finish this procedure, we have the following length N decomposition $w = [d_1, d_2, s_2]^T$. After the third iteration of the pyramid algorithm (once again, we apply filtering procedure to s_2), the N length decomposition $w = [d_1, d_2, d_3, s_3]^T$ is obtained. This procedure may be repeated up to J times where $J = \log_2(N)$ and gives the vector of wavelet

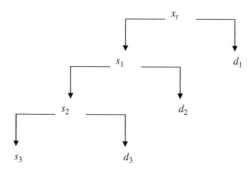

Fig. 1.9.   Analysis of a time series by a wavelet decomposition tree.
Note: This figure plots the wavelet decomposition tree. More specifically, the original time series can be decomposed into wavelet scaling coefficients and wavelet coefficients in the first step. In the next step, the scaling coefficients, obtained in the first step, is regarded as the original time series and decomposed as in the first step. This figure illustrates this procedure and continues to third step.

coefficients in Equation (1.32). This is called the wavelet decomposition tree and is presented in Fig. 1.9.

1.3.3. *Maximal overlap discrete wavelet transform*

We have examined the properties of the DWT as an alternative to the Fourier transform. In this sub-section, we examine the maximal overlap discrete wavelet transform (MODWT). It is natural to ask why the MODWT is needed instead of the DWT. The motivation for formulating the MODWT is essentially to define a transform that acts as much as possible like the DWT, but does not suffer from the DWT's sensitivity[5] to the choice of a starting point for a time series (Percival and Walden, 2000).

Non-redundancy of the DWT is achieved by down-sampling the filtered output at each scale (for detail, refer to Daubechies, 1992; Percival and Mofjeld, 1997). Importantly, the zero-phasing property of the MODWT permits meaningful interpretation of "timing" regarding the wavelet details. With this property, we can align perfectly the details from decomposition with the original time series. In comparison with the DWT, no phase shift will result in the MODWT.

The MODWT of level J for a time series x_t is a highly redundant non-orthogonal transform yielding the column vectors $\tilde{D}_1, \tilde{D}_2, \ldots, \tilde{D}_J$ and $\tilde{S}_J$, each of dimension N. The vector $\tilde{D}_j$ contains the MODWT coefficients

---

[5]This sensitivity results from down-sampling the outputs from the wavelet and scaling filters at each stage of the pyramid algorithm (Percival and Walden, 2000).

associated with changes in x_t between scale $j - 1$ and j, while $\tilde{S}_J$ contains the MODWT scaling coefficients associated with the smooth of x_t at scale J, or equivalently the variations of x_t at scale $J + 1$ and higher. The MODWT also follows the same pyramid algorithm as the DWT, while it utilizes the rescaled filters, instead of the wavelet and scaling filters in the DWT. These rescaled wavelet and scaling filters can be expressed as follows:

$$\tilde{h}_j = h_j/2^j \quad \text{and} \quad \tilde{g}_j = g_j/2^j \tag{1.37}$$

Utilizing its filtered output at each scale, a time series x_t can also be decomposed into its wavelet details and smooth as follows:

$$x_t = \sum_{j=1}^{J} \tilde{D}_j + \tilde{S}_J \tag{1.38}$$

However, this MRA of the MODWT provides some important features, which are not available to the original DWT. Percival and Walden (2000) present five important properties which distinguish the MODWT from the DWT:

(1) Although the DWT of level J restricts the sample size to an integer multiple of 2^J, the MODWT of level J is well defined for any sample size N.
(2) As in the DWT, the MODWT can be used to form an MRA. In contrast to the usual DWT, both the MODWT wavelet and scaling coefficients and the MRA are shift invariant in the sense that circularly shifting the time series by any amount will circularly shift by a corresponding amount the MODWT wavelet and scaling coefficients, details, and smooths. In other words, an MRA of the MODWT is associated with zero phase filters, implying that events, which feature in the original time series x_t, may be poorly aligned with features in the MRA.
(3) In contrast to the DWT details and smooths, the MODWT details and smooth are associated with zero phase filters, thus making it possible to meaningfully line up features in an MRA with the original time series x_t.
(4) As is true for the DWT, the MODWT can be used to form an analysis of variance based on the wavelet and scaling coefficients. However, the MODWT wavelet variance estimator is asymptotically more efficient than the same estimator based on the DWT.
(5) Whereas a time series and a circular shift of the series can have different DWT-based empirical power spectra, the corresponding MODWT-based spectra are the same.

To provide an example of the MODWT, we construct the MRA of the S&P500 returns. The data series are same as those of Fig. 1.7. The number of observations is 124. Note that with the MODWT, we are no longer limited to decomposing a sample size of dyadic length, i.e., a power of 2, while only limiting factor is the overall depth of the transform given by $J = \log_2(N) = 7$. However, in this example we choose a perform a level $J = 4$ MODWT on the return series using Haar and the Daubechies least asymmetric wavelet filter of length 8 [LA(8)] for the display purpose.

The left-hand side of Fig. 1.10 presents the MODWT coefficient vectors of r_t using Haar wavelet filter. Note that there are N wavelet coefficients at each scale because the MODWT does not down-sample after filtering.

Compared with Fig. 1.7, the MODWT coefficients are smoother than those of the DWT. This is because the MODWT wavelet coefficients $\tilde{d}_j$ contain the DWT coefficients, scaled by $1/\sqrt{2}$, and also the DWT

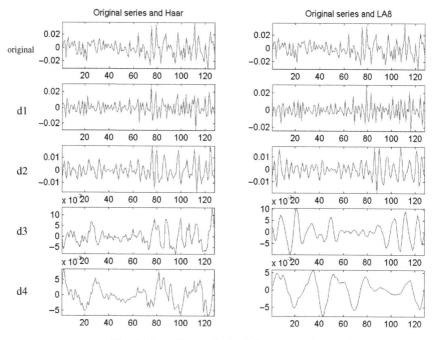

Fig. 1.10.   MRA analysis of the MODWT.

Note: This figure plots the MODWT-MAR using two different wavelet filters [Haar and LA(8)] from June 29, 2000 to December 29, 2000 using daily frequency. The left hand side of this figure is constructed using the Haar wavelet filter, while the right-hand of this figure is constructed by LA(8). From this figure, we can observe that the LA(8) wavelet filter generates more smooth wavelet coefficients than the Haar wavelet filter.

coefficients applied to x_t circularly shifted by one. Compared to the MODWT coefficients of the Haar wavelet, the coefficients filtered by LA(8) wavelet filter even more smoother. The longer wavelet filter has induced significant amounts of correlation between two adjacent coefficients, thus producing even smoother vectors of wavelet and scaling coefficients.

Practically, the wavelet coefficients are calculated using the wavelet filters. This introduces us another practical problem: how to choose a specific wavelet filter to implement wavelet analysis from the various wavelet filters. According to Gençay *et al.* (2010), to choose an appropriate wavelet filter, there are three aspects to be considered: length of data, complexity of the spectral density function, and the underlying shape of features in the data. First, the length of the original data is an important factor because the distribution of wavelet coefficients computed by the boundary conditions[6] will be very different from that of wavelet coefficients computed from complete sets of observations. The shorter the wavelet filter, the fewer wavelet coefficients produced.

Second, the complexity of the spectral density function has to be carefully considered to select a wavelet filter, since wavelet filters are finite in the time domain and thus infinite in the frequency domain. For example, if the spectral density function is quite dynamic, shorter wavelet filters may not be able to separate the activity between scales. In this case, longer wavelet filters would be more preferable to short wavelet filter. Clearly a balance between frequency localization and time localization is needed.

Finally, and most importantly, there is the issue of what the underlying features of the data look like. This is very important since wavelets are the basis functions of the data. If one chooses a wavelet filter that looks nothing like the underlying features, then the decomposition will be quite inefficient. Therefore, one should take care when selecting the wavelet filter and its corresponding basis function. Issues of smoothness and symmetry/asymmetry are the most common desirable characteristics for wavelet basis functions.

1.3.4. *Boundary condition*

In addition to the problem of choosing a proper wavelet filter, another condition should be considered when undertaking an empirical analysis using wavelet: the boundary condition. As indicated in choosing the wavelet

---

[6]The boundary conditions will be discussed more specifically at Section 1.3.4.

filter, the empirical data has a finite interval. This raises the issue of handling the boundaries.[7] In applying the DWT and the MODWT to finite length time series, there must be an established method for computing the remaining wavelet coefficients. Various techniques have been proposed to handle this problem, and three techniques in particular are briefly discussed in this section.

Periodic boundary

The most natural method for dealing with the boundary is to assume that the length N series is periodic, and to grab observations from the other end to finish the computations. In other words, any time series $f(x)$ defined on $[0, 1]$[8] could be expanded to live on the real line by regarding it as a periodic function with period one: $f(x) = f(x - [x])$ for $x \in R$. This is reasonable for some time series where strong seasonal effects are observed but cannot be applied universally in practice (Gençay *et al.*, 2002:144). This technique is generally adapted in wavelet analysis, partly because it is very easy to implement, and partly because the resulting empirical wavelet coefficients are independent with identical variances.

Reflection boundary

The reflection boundary condition is used extensively in Fourier analysis to reflect the time series about the boundaries. This technique produces a time series of length $2N$. This reflected series is applied to the wavelet transform under the assumption of periodic boundary conditions. More specifically, this technique consists of two methods: symmetric and antisymmetric reflection (see Ogden, 1997:112, for more detail). In a symmetric reflection, it is required to extend the domain of the function beyond $[0, 1]$ and define $f(x) = f(-x)$ for $x \in [-1, 0)$ and $f(x) = f(2 - x)$ for $x \in (1, 2]$. This has an advantage over the periodic boundary condition in the sense that it preserves the continuity of the function, though discontinuities in the derivatives of f may be introduced.

In addition, antisymmetric reflection causes the function to be reflected antisymmetrically about the endpoints. In terms of mathematical notations, $f(x) = 2f(0) - f(-x)$ for $x \in [-1, 0]$ and $f(x) = 2f(1) - f(2 - x)$ for

---

[7]Unser (1996) discusses some of the practical problems that arise when implementing these boundary conditions and argues that care must be given in coding the reconstruction algorithm to ensure that the original data can be recovered exactly.
[8]Since any interval can be translated, we consider here only the unit interval $[0, 1]$ without loss of generality.

$x \in (1, 2)$. This can preserve continuity in both the function and its first derivative. As with periodic boundary conditions, these methods impose their own alterations of the usual MRA. However, reflecting the time series does not alter the sample mean nor the sample variance, since all coefficients have been duplicated once (Gençay *et al.*, 2002).

Brick wall condition

Another way to handle the boundary is to impose the brick wall condition, which prohibits convolutions that extend beyond the ends of the series (Lindsay *et al.*, 1996). In other words, this condition can be implemented to simply remove any wavelet coefficient computed involving the boundary. Imposing this condition requires care when we calculate the wavelet variance and covariance.

As indicated in Lindsay *et al.* (1996), the brick wall condition can be used in an analysis, where data compression and regeneration is not the goals in which any convolution that extends beyond the end of the data series is not permitted. This boundary condition is appropriate in an analysis when there is no compelling reason to assume that the data are periodic and symmetric in structure.

1.4. Wavelet Variance, Covariance and Correlation

Variance, covariance and correlation are used to provide useful statistical information to researchers, and hence they are applied to many financial theories. In this section, we explain how the wavelet variance, covariance and correlation are derived in the wavelet domain.

1.4.1. *Wavelet variance*

In addition to the features of wavelet transforms (the DWT, the MODWT) stated in Sections 1.3.2 and 1.3.3, an important characteristic of wavelet transform is its ability to decompose or analyze the variance of a stochastic process. When we derive the wavelet coefficients using the DWT (or the MODWT), these wavelet coefficients indicate the changes at a particular scale. Using these coefficients, the wavelet variance on a particular scale can be obtained. In other words, the basic idea of the wavelet variance is to substitute the notion of variability over certain scales for the global measure of variability estimated by the sample variance (Percival and Walden, 2000). We first examine how the wavelet variances would be related to a sample variance. This can be seen by examining the sample variance of the time series y. The sample variance can be expressed as follows:

$$\hat{\sigma}_y^2 = \frac{1}{N} \sum_{i=1}^{N} [y_i - \bar{y}]^2$$

$$= \frac{1}{N} \sum_{i=1}^{N} [y_i]^2 - \bar{y}^2 \tag{1.39}$$

where $\bar{y}$ is the sample mean. Using orthogonality of the wavelet basis vectors, the sum of the series can be expressed as the sum of the squares of the wavelet coefficients.

$$\sum_{i=1}^{N} [y_i]^2 = \sum_{j=1}^{L} \sum_{k=1}^{n_j} d_{j,k}^2 + N\bar{y}^2 \tag{1.40}$$

Therefore, substituting Equation (1.40) into Equation (1.39) makes the variance a function of the wavelet coefficients.

$$\hat{\sigma}_y^2 = \frac{1}{N} \sum_{j=1}^{L} \left(\sum_{k=1}^{n_j} d_{j,k}^2 \right) \tag{1.41}$$

Let d and $\tilde{d}$ be the DWT and MODWT coefficient vectors, respectively. Equation (1.39) can be expressed as a vector notation under the assumption that a time series y has a zero mean.

$$\|y\|^2 = \|d\|^2 = \|\tilde{d}\|^2 \tag{1.42}$$

The relationship in Equation (1.42) provides a decomposition of variance between the original series and either the DWT or MODWT wavelet coefficients (Gençay *et al.*, 2002).

The wavelet variance for a time series X with a dyadic length $N = 2^J$ is estimated using the DWT coefficients for scale $\lambda_j \equiv 2^{j-1}$ through:

$$\hat{\sigma}_X^2(\lambda_j) \equiv \frac{1}{\hat{N}_j} \sum_{t=L'_j}^{N} [d_{j,t}^X]^2 \tag{1.43}$$

where $\hat{N}_j = N/2^j - L'_j$ is the number of wavelet coefficients at scale λ_j unaffected by the boundary[9] and $L'_j = [(L-2)(1-2^{-j})]$ is the number of the DWT coefficients computed using the boundary. While the spectral density

---

[9]How the boundary condition (especially the brick wall condition) affects the number of wavelet coefficients is explained in Section 1.3.4.

function decomposes the variance on a frequency-by-frequency basis, the wavelet variance decomposes the variance of X_t on a scale-by-scale basis.

We denote the MODWT coefficients of $X_1, \ldots, X_N$ as $\tilde{d}_{j,t}$ for $j = 1, \ldots, J$ and $t = 1, \ldots, N/2^j$. Similar to the variance of the DWT coefficients, the wavelet variance estimated by the MODWT coefficients for scale λ_j is as follows:

$$\tilde{v}_X^2(\lambda_j) \equiv \frac{1}{\tilde{N}_j} \sum_{t=L_j}^{N} [\tilde{d}_{j,t}^X]^2 \qquad (1.44)$$

where $\tilde{N}_j = N - L_j + 1$ is the number of coefficients unaffected by the boundary, and $L_j = (2^j - 1)(L - 1) + 1$ is the length of the scale λ_j wavelet filter. The decomposition of a time series as a sum of wavelet variances indicates which scales are important contributors to the time series variance (Percival and Walden, 2000; Serroukh and Walden, 2000). Percival (1995) and Percival and Walden (2000:309) provide the asymptotic relative efficiencies for the wavelet variance estimator based on the orthogonal DWT compared to the estimator based on the MODWT using a variety of power law processes. In their studies, they find that the DWT-based estimator can be rather inefficient — in the worst case, its large sample variance is twice that of the MODWT-based estimator.

To this point, we have examined how to derive the wavelet variances. For statistical inference, the confidence interval for the wavelet variance is required. Percival (1995) develops a theory for determining the uncertainty in the wavelet variance estimate for wavelet filters of various lengths under a Gaussian assumption. Under the assumption that the estimates of the wavelet variance of the DWT and the MODWT are unbiased and asymptotically normally distributed[10] (Lindsay *et al.*, 1996), the approximate $100(1 - 2p)\%$ confidence interval for the DWT estimate, $\hat{\sigma}_X^2(\lambda_j)$ and the MODWT estimate, $\tilde{v}_X^2(\lambda_j)$ can be derived:

$$\lfloor \hat{\sigma}_X^2(\lambda_j) - \Phi^{-1}(1 - p)\sqrt{\mathrm{var}(\hat{\sigma}_X^2(\lambda_j))},$$
$$\hat{\sigma}_X^2(\lambda_j) + \Phi^{-1}(1 - p)\sqrt{\mathrm{var}(\hat{\sigma}_X^2(\lambda_j))} \rfloor \qquad (1.45)$$

---

[10]Serroukh, Walden and Percival (2000) prove the wavelet variance is asymptotically normally distributed in three cases of the original time series: non-linear processes, non-Gaussian linear process, and non-stationary processes and differencing.

$$\lfloor \tilde{v}_X^2(\lambda_j) - \Phi^{-1}(1-p)\sqrt{\mathrm{var}(\tilde{v}_X^2(\lambda_j))},$$

$$\tilde{v}_X^2(\lambda_j) + \Phi^{-1}(1-p)\sqrt{\mathrm{var}(\tilde{v}_X^2(\lambda_j))}\rfloor \tag{1.46}$$

where $\Phi^{-1}(1-p)$ is the $(1-p) \times 100\%$ point for the standard normal distribution.

1.4.2. *Wavelet covariance and correlation*

In many economic and financial analyses, the temporal structure of the covariance between two series is of interest. This covariance structure can be applied to the wavelet analysis. The wavelet covariance is firstly compared to the Fourier cross spectra by Hudgins *et al.* (1993) using atmospheric surface-layer measurements of the horizontal and vertical velocities and the vertical velocity and temperature. In finance literature, the calculation of the wavelet covariance is a relatively new technique. Only a few researchers adopt this technique (see Gençay *et al.*, 2001, 2003, 2005; In and Kim, 2006).

The wavelet analysis of univariate time series can be generalized to multiple time series by defining the concept of the wavelet covariance between X_t and Y_t. As in standard statistics, the wavelet covariance can be defined as the covariance between the wavelet coefficients of X_t and Y_t at scale λ_j.

The sample covariance between X_t and Y_t is:

$$\sigma_{XY} = \frac{1}{N} \sum_{i=1}^{N} (X_{i,t} - \bar{X})(Y_{i,t} - \bar{Y})$$

$$= \frac{1}{N} \sum_{i=1}^{N} X_{i,t} Y_{i,t} - \bar{X}\bar{Y} = \frac{\langle XY \rangle}{N} - \bar{X}\bar{Y} \tag{1.47}$$

The inner product of two vectors in the last equality can be expressed in terms of their wavelet decompositions as follows:

$$\langle XY \rangle = \left\langle \sum_{j=1}^{L} \sum_{k=1}^{n_j} d_{j,k}^{X} \psi_{j,k} + \bar{X}1, \sum_{j=1}^{L} \sum_{k=1}^{n_j} d_{j,k}^{Y} \psi_{j,k} + \bar{Y}1 \right\rangle \tag{1.48}$$

Therefore, using the orthogonality properties of the vectors $\psi_{j,k}$, the sample covariance of the series may be written in terms of the wavelet coefficients.

$$\hat{\sigma}_{XY} = \sum_{i=1}^{L} \left\{ \frac{1}{N} \sum_{k=1}^{n_j} d_{j,k}^X d_{j,k}^Y \right\} \qquad (1.49)$$

As with the wavelet variance for univariate time series, the wavelet covariance also decomposes the covariance between two stochastic processes on a scale-by-scale basis. The term in the bracket in Equation (1.49) indicates the contribution to the covariance associated with each scale λ_j. More specifically, we can express the wavelet covariance at scale λ_j as follows:

$$\hat{\sigma}_{XY,j} = \frac{1}{N} \sum_{k=1}^{n_j} d_{j,k}^X d_{j,k}^Y \qquad (1.50)$$

If the brick wall boundary conditions are imposed, i.e., if the wavelet coefficients affected by the boundary are removed, the DWT estimate for the wavelet covariance can be derived as follows, analogous to the wavelet variance:

$$\hat{\sigma}_{XY,j} = \frac{1}{2^j \hat{N}_j} \sum_{k=L_j}^{\widehat{N}_j} d_{j,k}^X d_{j,k}^Y \qquad (1.51)$$

The MODWT wavelet covariance can also be expressed in terms of the MODWT wavelet coefficients:

$$\tilde{\sigma}_{XY,j} = \frac{1}{2^j \tilde{N}_j} \sum_{k=L_j}^{\tilde{N}_j} \tilde{d}_{j,k}^X \tilde{d}_{j,k}^Y \qquad (1.52)$$

The MODWT method allows a more accurate determination of the covariance associated with each scale (Lindsay *et al.*, 1996). Note that the estimator does not include any coefficients that make explicit use of the periodic boundary conditions. We can construct a biased estimator of the wavelet covariance by simply including the MODWT wavelet coefficients affected by the boundary and renormalizing.

As shown in Equations (1.51) and (1.52), the wavelet decomposition of the covariance is determined by the product of the coefficients from the two decompositions performed separately. Lindsay *et al.* (1996) show that the MODWT estimator $\tilde{\sigma}_{XY,j}$ is asymptotically normally distributed with mean $\sigma_{XY,j} \equiv 2^{-j} E\{\tilde{d}_j^X \tilde{d}_j^Y\}$ and variance

$$\text{var}(\tilde{\sigma}_{XY,j}) = \frac{1}{2^{2j+1}\tilde{N}_j} \left[\int_{-1/2}^{1/2} |S_{d_j^X d_j^Y}(w)|^2 dw \right.$$

$$\left. + \int_{-1/2}^{1/2} (S_{d_j^X}(w) S_{d_j^Y}(w)) dw \right] \qquad (1.53)$$

where $S_x(w)$ indicates the power spectral density at the frequency w of variable x. Based on these findings, an approximate $100 \times (1 - 2p)\%$ the confidence interval for the wavelet covariance can be constructed as follows:

$$\lfloor \sigma_{XY,j} - \Phi^{-1}(1 - p)\sqrt{\text{var}(\sigma_{XY,j})},$$

$$\sigma_{XY,j} + \Phi^{-1}(1 - p)\sqrt{\text{var}(\sigma_{XY,j})} \rfloor \qquad (1.54)$$

where $\Phi^{-1}(1 - p)$ is the $(1 - p) \times 100\%$ point for the standard normal distribution.

Because it is well known that the covariance does not take into account the variation of the univariate time series, it is natural to introduce the concept of the wavelet correlation. Although the wavelet covariance decomposes the covariance between two stochastic processes on a scale-by-scale basis and indicates a comovement between two series to some extent, in some situations, it would be more informative to normalize the wavelet covariance by the variability calculated from the observed wavelet coefficients. Statistically, it is necessary to calculate the wavelet correlation. The wavelet correlation is simply made up of the wavelet covariance for $\{X_t, Y_t\}$, and wavelet variances for $\{X_t\}$ and $\{Y_t\}$. The wavelet correlation can be expressed as follows:

$$\tilde{\rho}_{XY}(\lambda_j) \equiv \frac{\text{Cov}_{XY}(\lambda_j)}{\tilde{v}_X(\lambda_j) \tilde{v}_Y(\lambda_j)} \qquad (1.55)$$

As with the usual correlation coefficient between two random variables, $|\tilde{\rho}_{XY}(\lambda_j)| < 1$. The wavelet correlation is analogous to its Fourier equivalent, the complex coherency (Gençay *et al.*, 2002:258).

We now turn our attention to the confidence interval of the wavelet correlation. Given the inherent non-normality of the correlation coefficient for small sample sizes, a non-linear transformation is sometimes required in order to construct a confidence interval. Let $h(\rho) \equiv \tanh^{-1}(\rho)$ define Fisher's z-transformation. For the estimated correlation coefficient $\hat{\rho}$, based on N independent samples, $\sqrt{N-3}[h(\hat{\rho}) - h(\rho)]$ is approximately distributed as a Gaussian with mean zero and unit variance. Based on these findings, an approximate $100 \times (1 - 2p)\%$ the confidence interval for the wavelet correlation can be constructed as follows:

$$\left[\tanh\left\{ h[\hat{\rho}_{XY}(\lambda_j)] - \frac{\Phi^{-1}(1-p)}{\sqrt{\hat{N}_j - 3}} \right\}, \tanh\left\{ h[\hat{\rho}_{XY}(\lambda_j)] + \frac{\Phi^{-1}(1-p)}{\sqrt{\hat{N}_j - 3}} \right\} \right]$$

(1.56)

where $\hat{N}_j$ is the number of wavelet coefficients associated with scale λ_j computed via the DWT — not the MODWT. This assumption of uncorrelated observations in order to use Fisher's z-transformation is only valid if we believe no systematic trends or non-stationary features exist in the wavelet coefficients at each scale.

1.4.3. *Wavelet cross covariance and correlation*

The cross correlation is a more powerful tool for examining the relationship between two time series. The cross correlation function considers the two series not only simultaneously (at lag 0), but also with a time shift. The cross correlation reveals causal relationships and information flow structures in the sense of Granger causality. If two time series were generated on the basis of a synchronous information flow, they would have a symmetric lagged correlation function, $\rho_\tau = \rho_{-\tau}$; the symmetry would be violated only by insignificantly small, purely stochastic deviations. As soon as the deviations between ρ_τ and $\rho_{-\tau}$ become significant, there is asymmetry in the information flow and a causal relationship that requires an explanation.

The cross correlation can be constructed utilizing the wavelet cross covariance. It is straightforward to derive the cross covariance, once the wavelet covariance is derived. For $N \geq L_j$, a biased estimator of the wavelet cross covariance based on the MODWT is given by:

$$R_{XY,\tau} = \begin{cases} \dfrac{1}{\tilde{N}} \displaystyle\sum_{t=L_j-1}^{N-\tau-1} \tilde{d}_{j,t}^X \tilde{d}_{j,t+\tau}^Y & \text{for } \tau = 0, \dots, \tilde{N}_j - 1 \\[3ex] \dfrac{1}{\tilde{N}} \displaystyle\sum_{t=L_j-1}^{N-\tau-1} \tilde{d}_{j,t}^X \tilde{d}_{j,t+\tau}^Y & \text{for } \tau = -1, \dots, -(\tilde{N}_j - 1) \\[3ex] 0 & \text{otherwise} \end{cases}$$

(1.57)

Allowing the two processes to differ by an integer lag τ, the wavelet cross correlation can be defined as:

$$\rho_{\tau,XY}(\lambda_j) \equiv \frac{R_{XY,\tau}(\lambda_j)}{\tilde{v}_X(\lambda_j)\tilde{v}_Y(\lambda_j)} \tag{1.58}$$

1.5. Long Memory Estimation Using Wavelet Analysis

Recently, several works have found evidence of stochastic long memory behavior in the financial time series. The presence of long memory dynamics, which is a special form of non-linear relationships, indicates non-linear dependence in the first moment of the distribution, and hence provides a potentially predictable component in the series dynamics. In this section, we describe and summarize the estimation procedure of the long memory parameter using wavelet analysis, based on the studies of Jensen (1999a, 1999b, 2000) and Manimaran *et al.* (2005, 2006). More specifically, in Section 1.5.1, the definition of long memory and the meaning of long memory parameter are described. We present the wavelet ordinary square (WOLS) in Section 1.5.2, while in Section 1.5.3, we explain how to derive the maximum-likelihood estimator for the long memory parameter. Another method for estimating long memory parameter will be discussed in Section 1.5.4.

1.5.1. *Definitions of long memory*

There are several possible definitions of the property of long memory. According to McLeod and Hipel (1978), a discrete time series x_t, with autocorrelation function, ρ_j at lag j, possesses long memory if the quantity

$$\lim_{n\to\infty} \sum_{j=-n}^{n} |\rho_j| \tag{1.59}$$

is non-finite. Equivalently, the spectral density $S(w)$ will be unbounded at low frequencies. A stationary and invertible ARMA process has autocorrelations, which are geometrically bounded, and hence is a short memory process (Baillie, 1996). Fractionally integrated processes are long memory processes given the definition in Equation (1.59).

Let x_t denote a fractionally integrated process, $I(d)$, defined by:

$$\begin{aligned} (1-L)^d x_t &= e_t \\ x_t &= (1-L)^{-d} e_t \end{aligned} \tag{1.60}$$

where L is the lag operator, $e_t \sim i.i.d.N(0, \sigma_e^2)$, and d is the fractional differencing parameter, which is allowed to assume any real value in $(0, 1)$. The process x_t is covariance-stationary for $0 < d < 1/2$, but not otherwise.[11] While the process x_t is covariance-stationary when $0 < d < 1/2$, its autocovariance function declines hyperbolically to zero, making x_t a long-memory process. If $1/2 < d < 1$, x_t has an infinite variance, but still has a mean-reverting property in the very long run.

1.5.2. *Wavelet ordinary least square*

As shown in Jensen (1999b), the wavelet coefficient from an $I(d)$ process has a variance that is a function of the scaling parameter, j, but is independent of the translation parameter, k. McCoy and Walden (1996) and Jensen (1999b) demonstrate that wavelet coefficients have a normal distribution with zero mean and variance, $\sigma^2 2^{-2jd}$. Taking logarithms on the wavelet coefficient's variance yields the following relationship.

$$\ln R(j) = \ln \sigma^2 - d \ln 2^{2j} \tag{1.61}$$

where $R(j)$ denotes the wavelet coefficient's variance and is linearly related to $\ln 2^{-2j}$ by the fractional differencing parameter, d. However, note that due to the restriction of the DWT (discrete wavelet transform), the number of observations for the underlying process, x_t, must be a power of 2. Using Monte Carlo experiments, Jensen (1999b) demonstrates that the small and large properties of the wavelet OLS estimator are superior to the GPH estimator. More specifically, the WOLS estimator has a lower MSE (mean squared errors) than the GPH estimator.

1.5.3. *Approximate maximum-likelihood estimation of the long memory parameter*

Wavelet-based maximum likelihood estimation procedures, related to economic and finance research, have been studied by Jensen (1999a, 2000). Although least squares estimation is popular because of its simplicity to program and compute, it produces much larger mean square errors when compared to maximum likelihood methods. Another advantage of the wavelet-based MLE is that the long memory estimator, d, is unaffected

---

[11]When $d < 1$, the process is called "mean reverting", although this terminology needs to be used with care, since the existence of the mean is not easily shown when the variance is undefined.

by the unknown μ, since the wavelet coefficients autocovariance function is invariant to μ (Jensen, 2000). The approximate maximum likelihood methodology,[12] proposed in Jensen (1999a, 2000), overcomes the difficulty of computing the exact likelihood by replacing the covariance matrix of the process with an approximation using the DWT.[13] This is possible through the ability of the DWT to decorrelate the long memory process.

If x_t is a length $N = 2^J$ FDP with mean zero and covariance matrix given by Ω_x, then the likelihood can be expressed as follow (see Gençay *et al.*, 2002:172):

$$L(d, \sigma_\varepsilon^2 | x) = (2\pi)^{-N/2} |\Omega_x|^{-1/2} \exp\left(-\frac{1}{2} x^T \Omega_x^{-1} x\right) \qquad (1.62)$$

where the quantity $|\Omega_x|$ is the determinant of Ω_x. The maximum-likelihood estimators (MLEs) of the parameters (d and σ_ε^2) are those quantities that maximize Equation (1.62). As in Gençay *et al.* (2002), to avoid the difficulties in computing the exact MLEs, we use the approximation of the DWT as applied to FDPs. In other words, the covariance matrix Ω_x is expressed by:

$$\Omega_x \approx \tilde{\Omega}_x = W^T \Sigma_x W \qquad (1.63)$$

where W is the orthogonal matrix defining the DWT and Σ_x is a diagonal matrix containing the variances of DWT coefficients. Using Equation (1.62), we try to find the values of d and σ_ε^2 that minimize the following log-likelihood function.

$$\tilde{L}(d, \sigma_\varepsilon^2 | x) = -2\log(\tilde{L}(d, \sigma_\varepsilon^2 | x)) - N\log(2\pi) = \log(|\tilde{\Omega}_x|) + x^T \Omega_x^{-1} x \qquad (1.64)$$

1.5.4. *Another estimation method of the long memory parameter*

Using this wavelet variance (wavelet power), Manimaran *et al.* (2005, 2006) calculate the fluctuation function $F(\lambda_j)$ at a scale λ_j, where the fluctuation is extracted by subtracting the reconstructed series after removal of the

---

[12]Cheung and Diebold (1994) find that the approximate MLE can be an efficient and attractive alternative to the exact MLE when μ is unknown.
[13]The wavelet MLE enjoys the advantage of having both the strengths of an MLE and a semiparametric estimator, but does not suffer their known drawbacks (Jensen, 2000).

successive wavelet coefficients from the data, as follows:

$$F(\lambda_j) = \left[\sum_{\lambda_j=1}^{s} \sigma^2(\lambda_j) \right]^{1/2} \tag{1.65}$$

This fluctuation function $F(\lambda_j)$ can be used to obtain the scaling behavior of a time series through:

$$F(\lambda_j) \sim \lambda_j^H \tag{1.66}$$

where H is the Hurst exponent scaling exponent, which can be obtained from the slope of log-log plot of $F(\lambda_j)$ versus scale λ_j. The Hurst exponent equal to 0.5 for the Gaussian white noise. $H > 1/2$ is for persistent time series, and $H < 1/2$ for anti-persistent (anti-correlated) time series.

Chapter 2

Multiscale Hedge Ratio Between the Stock and Futures Markets: A New Approach Using Wavelet Analysis and High Frequency Data

In this chapter, we investigate the multiscale relationship between the stock and futures markets over various time horizons. We propose a new approach — the wavelet multiscaling method — to undertake this investigation. This method enables us to make the *first* analysis of the multiscale hedge ratio using high frequency data (5 min). Wavelets are treated as a "lens" that enables the researcher to explore the relationships that previously were unobservable. The approach focuses on the relationship in three ways: (1) the lead-lag causal relationship, (2) covariance/correlation, and (3) the hedge ratio and hedging effectiveness. Our empirical results show that the future market Granger causes the stock market. We find that the magnitude of the wavelet correlation between the two markets increases as the time scale increases, indicating that the stock and futures markets are not fundamentally different. We also find that the hedge ratio at the second scale has the lowest value and increases monotonically at a decreasing rate, converging toward the long horizon hedge ratio of one, which suggests that the shared permanent component ties the stock and futures series together, and the effect of the transitory components becomes negligible.

2.1. Introduction

Many of the participants in futures markets are hedgers. Hedging is one of the important reasons for using derivative securities such as futures contracts. The literature shows that the study of the hedge ratio using futures contracts has been of interest to both academicians and practitioners and widely discussed. In addition to this aspect, understanding the long-run and short-run relationships between the stock and futures markets is important for portfolio management. For example, in financial risk management, risk is assessed at different time scales, which vary from intervals as small as a few minutes to longer time scales such as days or even months. A common practice in the risk management industry is the conversion of short-scale risk measures into longer horizons by taking the corresponding scaling quantity into account. Based on this practice, we would question whether the long-run covariance/correlation between the stock and futures markets is similar to that of the short-run. We can then question whether the long-run hedge ratio is similar to the short-run hedge ratio. As in the study of Lee (1999), given the time-varying nature of the covariance in many financial markets, the classical assumption of a time-invariant optimal hedge ratio[1] appears inappropriate.

To determine the optimal hedge ratio, some early studies assume that the hedge ratio is constant over time and estimate it using simple ordinary least square estimation (see Ederington, 1979; Hill and Schneeweis, 1982). However, given the time-varying nature of the covariance in many financial markets, the classical assumption of the time-invariant optimal hedge ratio appears inappropriate. An improvement has been made by adopting a bivariate GARCH framework (see Kroner and Sultan, 1993; Choudhry, 2003; Wang and Low, 2003).

Traders and investors working in the stock and futures markets have a different hedging horizon. Therefore, examining and applying the one-period hedge ratio could lead investors to invalid decision making. The previous literature has shown little attention to the multiscale hedge ratio, except for Howard and D'Antonio (1991), Lien and Luo (1993, 1994), Geppert (1995), and Lien and Wilson (2001). However, the models presented by these authors have at least three problems in estimating the multiscale hedge ratio. First, the ratio is an unreliable estimator due to a handful of independent observations generated from long-horizon return series (see Geppert, 1995). Second, computation is burdensome and difficult to calculate over longer investment horizons. Finally, it requires an assumption for the error term for GARCH/SV model estimation (see Lien and Wilson, 2001), which can cause inaccurate results.

---

[1]Since risk in this context is usually measured as the volatility of portfolio returns, an intuitively plausible strategy might be to choose the hedge ratio that minimizes the variance of the returns of a portfolio containing the stock and futures position. This is known as the optimal hedge ratio (Brooks *et al.*, 2002).

To overcome these problems, In and Kim (2006) recently proposed a new approach for investigating the relationship between the stock and futures markets using wavelet analysis. The main advantage of using wavelet analysis is the ability to decompose the financial data into several time scales (investment horizons). All market participants such as regulators and speculative investors who trade in the stock and futures markets make decisions over different time scales. The time frame over which they operate differs immensely, ranging from seconds to months and beyond. In fact, due to the different decision-making time scales among traders, the true dynamic structure of the relationship between the stock and futures markets itself will vary over the different time scales associated with those different horizons. Although it is logical to assume that there are several time periods in decision making, economic and financial analyses have been restricted to at most two time scales (the short-run and the long-run), due to the lack of analytical tools to decompose data into more than two time scales. However, unlike previous studies, this chapter uses wavelets to produce an orthogonal decomposition of correlation and the hedge ratio between the stock and futures indices over several different time scales (In and Kim, 2006).

Recently, several applications of wavelet analysis to economics and finance have been documented in the literature. To the best of our knowledge, applications in these fields include examination of foreign exchange data using waveform dictionaries (Ramsey and Zhang, 1997), decomposition of economic relationships of expenditure and income (Ramsey and Lampart, 1998a, 1998b), the multihorizon Sharpe ratio (Kim and In, 2005a), systematic risk in a capital asset pricing model (Gençay *et al.*, 2003), and examination of the multiscale relationship between stock returns and inflation (Kim and In, 2005b).

This chapter is different from the previous studies. We utilize intraday data. Adopting intra-day data and wavelet analysis is very useful for examining the multiscale hedge ratio in that it allows us to investigate how the hedge ratio can be affected by hedge horizon from intra-day to longer horizons. More specifically, we can observe the difference between the 5-minute hedge ratio and the much longer horizon hedge ratio.

Our empirical results indicate that first, at 5-minute (original data) and approximately one-day dynamics (seventh scale), the futures market Granger causes the stock market. From this result, it is concluded that the futures market is more efficient. Examining the wavelet variance reveals an approximate decreasing linear relationship between the wavelet variance and the wavelet scale. This implies that an investor with a short investment horizon has to respond to every fluctuation in realized returns, while the long-run risk for an investor with a longer horizon is significantly less. We find that the magnitude of the wavelet correlation between the two markets increases as the time scale increases, indicating that the stock and futures markets are not fundamentally different. Finally, we also find that the hedge ratio at the second scale has the lowest value and increases monotonically at a decreasing rate, converging toward the long horizon hedge ratio of one, which suggests that the shared permanent component ties the stock and futures series together and the effect of the transitory components becomes negligible.

This chapter is organized as follows. Section 2.2 derives the minimum variance hedge ratio. Section 2.3 discusses the data and the empirical results. In Section 2.4, concluding remarks are presented.

2.2. Minimum Variance Hedge

The most widely used static hedge ratio[2] is the minimum variance (MV) hedge ratio, which is derived Johnson (1960) by minimizing the portfolio risk. In this framework, the risk is given by the variance of changes in the value of the hedged portfolio. Assume that an individual has taken a fixed position in some asset and that this person is long one unit of the asset without loss of generality. Let h_t represent the short position taken in the futures market at time t under the adopted hedging strategy. Ignoring daily resettlement, the hedger's objective within this framework is to minimize the variance of the change in the value of the hedged portfolio:

$$\text{Min Var}(\Delta HP_t)$$

$$= \text{Var}(\Delta S_t + h_t \Delta F_t)$$

$$= \text{Var}(\Delta S_t) + h_t^2 \text{Var}(\Delta F_t) + 2h_t \text{Cov}(\Delta S_t, \Delta F_t) \qquad (2.1)$$

where ΔHP_t is the change in the value of the hedged portfolio during time t; ΔS_t and ΔF_t are the changes in the log of the stock and the futures prices at time t, respectively; and h_t is the optimal hedge ratio. Duffie (1989) shows that the optimal hedge ratio for a person with mean-variance utility can be decomposed into two portions: one reflecting speculative

---

[2]We use the static MV hedge ratio, which changes with the holding period, not with time. Our purpose is not to examine the time-varying hedge ratio, but to investigate the horizon-varying hedge ratio.

demand (which varies across individuals according to their risk aversion) and another reflecting a pure hedge (which is the same for all mean-variance utility hedgers). Because the pure hedge term is common to all hedgers, and the speculative demand term is both difficult to estimate and often close to zero, it is reasonable to focus attention on the pure hedge.

Suppose the hedger decides to pursue a hedging strategy. The optimal hedge is determined by solving Equation (2.1).

$$\frac{\partial \text{Var}(\Delta HP_t)}{\partial h_t} = 2h_t \text{Var}(\Delta F_t) + 2\text{Cov}(\Delta S_t, \Delta F_t)$$

$$h_t^* = -\frac{\text{Cov}(\Delta S_t, \Delta F_t)}{\text{Var}(\Delta F_t)} = \rho_{sf}\frac{\sigma_s}{\sigma_f} \qquad (2.2)$$

where ρ_{sf} is the correlation coefficient between ΔS_t and ΔF_t, and σ_s and σ_f are standard deviations of ΔS_t and ΔF_t. This corresponds to the conventional hedge ratio, when changes in both stock and futures prices are homoskedastic. In the absence of conditional heteroskedasticity, both $\text{Cov}(\Delta S_t, \Delta F_t)$ and $\text{Var}(\Delta F_t)$ are independent of the information set. As a result, h_t^* is a constant term regardless of whatever information is available. Commonly, the value of the hedge ratio is less than unity, so that the hedge ratio that minimizes risk in the absence of basis risk turns out to be dominated by h_t^* when basis risk is taken into consideration (Choudhry, 2003). The attractive feature of the MV hedge ratio is that it is easy to understand and simple to compute. In general, it is considered that the MV hedge ratio is not consistent with the mean-variance framework, since it ignores the expected return on the hedged portfolio. However, for the MV hedge ratio to be consistent with the mean-variance framework, either the investors need to have an infinite risk aversion coefficient or the expected return on the futures contracts is zero, in other words, the futures price follows a pure martingale process.[3]

The degree of hedging effectiveness, proposed by Ederington (1979), is measured by the percentage reduction in the variance of the stock prices changes. Therefore, the degree of hedging effectiveness, denoted as *EH*, can

---

[3]Other strategies that incorporate both the expected return and risk (variance) of the hedged portfolio have been recently proposed: Optimum mean-variance hedge ratio by Hsin *et al.* (1994), Sharpe hedge ratio by Howard and D'Antonio (1984), and minimum generalized semivariance (GSV) hedge ratio suggested by Chen *et al.* (2001). However, it is shown that if the futures price follows a pure martingale process or if the futures and spot returns are jointly normally distributed, the optimal mean-variance hedge ratio will be the same as the MV hedge ratio.

be expressed as follows:

$$EH = \frac{\text{Var}(\Delta S_t) - \text{Var}(\Delta HP_t)}{\text{Var}(\Delta S_t)} = 1 - \frac{\text{Var}(\Delta HP_t)}{\text{Var}(\Delta S_t)} = \rho_{sf}^2 \qquad (2.3)$$

where $\rho_{sf,t}^2$ is the square of the correlation between the change in the stock and futures prices.

Given the wavelet variance and covariance between two series, the hedge ratio at scale λ_j can be calculated using Equations (1.44) and (1.52).

$$h_j^w = \frac{\text{Cov}_{sf}(\lambda_j)}{\tilde{v}_f^2(\lambda_j)} \qquad (2.4)$$

In this specification, h_j^w indicates the wavelet multiscale hedge ratio, which can vary depending on the wavelet scales (i.e., investment horizons). To understand how the wavelet scale represents the investment horizon, it is more informative to consider the spectral representation theorem. By this theorem, the spectrum of original 5-minute return series, R_t, contains all frequencies between zero and 1/2 cycles, equivalent to a 0–10 minute period in our data frequency. The wavelet coefficients at the first scale, d_1, are associated with the frequencies in the interval [1/4, 1/2], equivalent to a 10–20 minute period, while the wavelet coefficients at the second and third scales, d_2 and d_3, respectively, contain frequencies [1/8, 1/4] and [1/16, 1/8], respectively. Thus, the hedge ratios calculated at different scales represent the short position taken in the futures market at various frequencies (in other words, at various time scales).

2.3. Empirical Results

In this chapter, we use the S&P500 Index and the S&P500 Futures Index maturing in June 2004. Our data is composed of 5-minute prices for each index obtained from Tickdata (www.tickdata.com) for the period March 22, 2004 to June 9, 2004. The final three-months' trading data has been chosen for our study because most trading occurs in and close to the expiry month. The trading hours of the Chicago Mercantile Exchange (CME) are 8:30 to 15:15 local time, while the New York Stock Exchange (NYSE) opens from 9:30 to 16:00 local time. However, the local time of New York is one hour ahead of Chicago. Therefore, the opening hours of both markets are the same, while the CME closes 45 minutes earlier than the NYSE. Therefore, in our study we use the time period from 8:30 to 15:00, which generates 78 observations per day. In our sample we have 56 trading days, which

Table 2.1. Basic statistics.

	Futures market	Stock market
Mean	0.0003	0.0003
Variance	0.0016	0.0014
Skewness	0.2966	0.4281
Kurtosis	14.8162	13.5827
JB	39979.8616	33679.8516
	(0.0000)	(0.0000)
ρ	−0.0374	0.0129
LB(15) for R_t	22.2503	26.8247
	(0.0348)	(0.0082)
ρ^2	0.0075	0.0119
LB(15) for R_t^2	31.7695	55.2397
	(0.0015)	(0.0000)

Note: Sample period: March 22, 2004 – June 9, 2004. The means and variances are calculated by multiplying by 100 and 10,000, respectively. Significance levels are in parentheses. LB(n) is the Ljung-Box statistic for up to n lags, distributed as χ^2 with n degrees of freedom. ρ and ρ^2 indicate the first-order autocorrelations of returns and squared returns, respectively. Skewness and kurtosis are defined as $E[(R_t - \mu)]^3$ and $E[(R_t - \mu)]^4$, where μ is the sample mean. JB indicates the Jarque–Bera statistic.

generates a total of 4368 (56×78) observations. The 5-minute changes of both stock and futures indexes are calculated by $\log(P_t) - \log(P_{t-1})$.

Table 2.1 summarizes selected basic statistics. All sample means are positive in the sample period. Variances are 0.0016 for the futures index and 0.0014 for the stock index, showing that the futures market has a higher volatility than the stock market. The signs of skewness are all positive. The values of Ljung-Box up to 15 lags (hereafter LB(15)) for the return series are significant at the 5% level. The LB(15) for squared return series are highly significant for both markets, suggesting the possibility of the presence of autoregressive conditional heteroskedasticity. First-order autocorrelations of futures and stock returns (ρ) are −0.0374 and 0.0129, respectively. For the squared return data, the first-order serial correlations (ρ^2) for the futures and stock markets are 0.0075 and 0.0119, respectively, indicating that the stock returns are more persistent than the futures returns.

To analyze the relationship between the stock and the futures markets using wavelet analysis, we need to choose which wavelet filter to use from the various wavelet filters. In consideration of the balance between the sample size and the length of wavelet filter, we settle with the Daubechies least asymmetric wavelet filter of length 8 [LA(8)],[4] while we decompose our data up to scale 9. To examine the lead-lag relationship in wavelet analysis, first, we test for Granger causality up to level 9, including the original data set.

The results of the Granger causality tests are reported in Table 2.2. As can be seen in Table 2.2, the futures market leads the stock market at the original data and the seventh scale, while at the other scales, there are feedback relationships. Note that lower scales correspond to higher frequency bands. For example, the first scale is associated with 5-minute changes, the second scale is associated with $2 \times 5 = 10$-minute changes, and so on.

The first six scales capture the frequencies $1/64 \leq f \leq 1/2$; i.e., oscillations with a period length of 2 (10 minutes) to 64 (320 minutes). Since there are $78 \times 5 = 390$ minutes in a day, the first six scales relate to the intra-day dynamics of our sample. This is of interest in that at 5-minute (original data) and approximately one-day dynamics (seventh scale), the futures market Granger causes the stock market. The cost-of-carry (COC) model states that as new information arrives simultaneously to the stock and futures markets and is reflected immediately in both the stock and futures prices, profitable arbitrage should therefore not exist, under the assumption that the two markets are perfectly efficient and frictionless and act as perfect substitutes. In other words, if both markets are efficient, the COC model indicates that the two markets have a feedback relationship in terms of Granger causality. Based on this finding of the Granger causality test, we can conclude that the futures market is more efficient. In other words, price discovery is greater in the futures market (Lin and Stevenson, 2001).

Turning to the second purpose of our chapter (correlation in the various time scales), we first examine the variances of the futures and the stock markets' returns in various time scales. An important characteristic of the wavelet transform is its ability to decompose (analyze) the variance of the stochastic process. Figs. 2.1(a) and 2.1(b) illustrate the MODWT-based wavelet variances of two series in log-log scales. The straight lines indicate the variance and the dotted lines indicate the 95% confidence interval.[5]

---

[4]In our study, to check the appropriateness of our choice of wavelet filter, we examined the same test using other wavelet filters, such as the Daubechies extremal phase wavelet filter of length 4, D(4) and the Daubechies least asymmetric wavelet filter of length 16, D(16). The results are not quantitatively different from those of LA(8). To conserve space, the results are not reported but available from the authors on request.

[5]For a detailed explanation on how to construct the confidence interval of wavelet variance, see Gençay *et al.* (2002:242).

Table 2.2.   Granger causality test.

	Original	1	2	4	8	16	32	64	128	256
Stock → Futures	1.83	62.46*	108.30*	382.27*	38.34*	141.66*	21.10*	1.22	49.92*	77.29*
	(0.12)	(0.00)	(0.00)	(0.00)	(0.00)	(0.00)	(0.00)	(0.30)	(0.00)	(0.00)
Futures → Stock	818.80*	579.75*	431.55*	240.74*	435.70*	603.02*	631.01*	557.13*	514.61*	204.49*
	(0.00)	(0.00)	(0.00)	(0.00)	(0.00)	(0.00)	(0.00)	(0.00)	(0.00)	(0.00)

Note: The original data has been transformed by the wavelet filter (LA(8)) up to time scale 9. * indicates significance at 5% level. The significance levels are in parentheses. The first detail (wavelet coefficient) d1 captures oscillations with a period length 5 to 10 min. The last wavelet scale captures oscillations with a period length of 1280 to 2560 min.

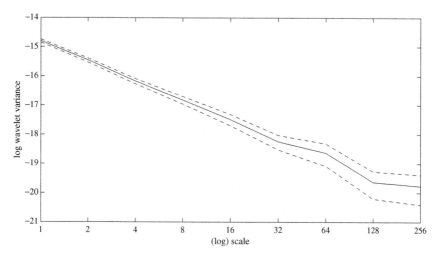

Fig. 2.1a.   Estimated wavelet variance of stock returns.
Note: Wavelet variances for stock returns are plotted on different time scale x-axis. The dotted lines represent approximate 95% confidence interval. Each scale is associated with a particular time period. For example, the first scale is 5 min, the second scale is $2 \times 5 = 10$ min, the third scale is $4 \times 5 = 20$ min and so on. The seventh scale is $64 \times 5 = 320$ min. Since there are 390 min per day, the seventh scale corresponds to approximately one day.

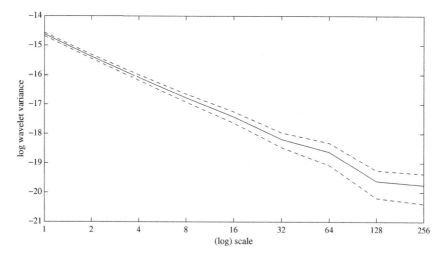

Fig. 2.1b.   Estimated wavelet variance of futures returns.
Note: Wavelet variances for futures returns are plotted on different time scale x-axis. The dotted lines represent approximate 95% confidence interval. Each scale is associated with a particular time period. For example, the first scale is 5 min, the second scale is $2 \times 5 = 10$ min, the third scale is $4 \times 5 = 20$ min and so on. The seventh scale is $64 \times 5 = 320$ min. Since there are 390 min per day, the seventh scale corresponds to approximately one day.

There is an approximate linear relationship between the wavelet variance and the wavelet scale, indicating the potential for long memory in the volatility series. The variances of both the stock and futures markets decrease as the wavelet scale increases. Note that the variance-versus-wavelet scale curves show a broad peak at the first scale in both markets. More specifically, a wavelet variance in a particular time scale indicates the contribution to sample variance (Lindsay *et al.*, 1996:778). The sample variances of the stock and futures markets are 0.0014 and 0.0016, respectively, and 49%[6] and 51% of the total variances of the stock and futures markets, respectively, are accounted for by the first scale. This result implies that an investor with a short investment horizon has to respond to every fluctuation in the realized returns, while for an investor with a much longer horizon, the long-run risk is significantly less. In addition, notice that the wavelet variances show that the futures market is more volatile than the stock market regardless of the time scale. This is consistent with the results of Lee (1999), who finds that the futures market has higher volatility than the stock market using a GARCH model.

Furthermore, the eighth scale, where we observe an apparent break in the variance for both series, is associated with $128 \times 5 = 640$-minute changes. Since there are 390 minute in one day, the eighth scale corresponds to 1.64 days. This implies that the stock and futures returns have a multiscaling behavior. There are two different scaling parameters corresponding to intra-day and higher scales, respectively. This apparent multiscaling behavior occurs approximately at the daily horizon where the intra-day seasonality is strongly present. The removal of the intra-day seasonality would not eliminate this multiscaling, but the transition between two scaling regions would be more gradual by exhibiting a concave scaling behavior (Gençay *et al.*, 2001).

In addition to examining the variances of two time series, a natural question is to consider how the two series are associated with one another. The wavelet correlation is constructed to examine how the two series are related over various time scales. Fig. 2.2 illustrates the estimated wavelet correlation between the stock and futures returns against the various time scales. In this figure, the straight line indicates the wavelet correlations up to ninth scale, while the dotted lines represent the 95% confidence interval.

---

[6]This figure can be calculated by the normalization of the wavelet variance using the sample variance. For more detail, see Lindsay *et al.* (1996).

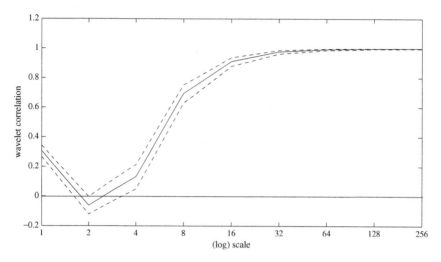

Fig. 2.2.   Estimated wavelet correlation between stock and futures returns.
Note: The dotted lines denote approximate 95% confidence interval. At scales d1–d3, correlation rapidly increases and at scales d4–d8, gradual and persistent behavior becomes more visible. Each scale is associated with a particular time period. For example, the first scale is 5 min, the second scale is $2 \times 5 = 10$ min, the third scale is $4 \times 5 = 20$ min and so on. The seventh scale is $64 \times 5 = 320$ min. Since there are 390 min per day, the seventh scale corresponds to approximately one day.

Two things are worth noting in Fig. 2.2. First, the correlations up to the seventh scale are increasing, starting from the second scale. In the intra-day scales (up to seventh scale), the correlation coefficients are very different, depending on the specific time scale, while at one-day and higher dynamics, the correlation remains very high, close to one. This implies that in the intra-day scales, especially at very short scales, the primary purpose for traders' participation in the futures market is not hedging, but speculation. Second, the magnitude of the correlation increases as the time scale increases, indicating that the stock and futures markets are not fundamentally different (Lee, 1999).

The final purpose of this chapter is to examine the multiscale hedge ratio, based on the results of variance and covariance obtained from wavelet analysis. As indicated in Lien and Luo (1993, 1994), realism suggests that the hedger's planning horizon usually covers multiple periods. Therefore, examining the multiscale hedge ratio is more appropriate than examining the one-period hedge ratio. Fig. 2.3 shows the hedge ratio and hedging effectiveness using the various time scales up to scale 9. Note that these wavelet hedge ratios are estimated by a nonparametric method. Therefore,

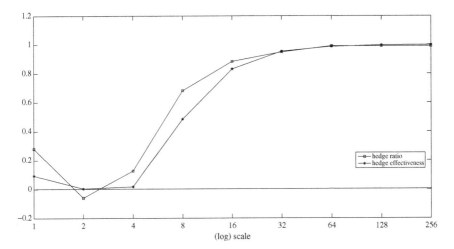

Fig. 2.3.   Hedge ratio and hedging effectiveness against wavelet domains.
Note: Each scale is associated with a particular time period. For example, the first scale is 5 min, the second scale is $2 \times 5 = 10$ min, the third scale is $4 \times 5 = 20$ min and so on. The seventh scale is $64 \times 5 = 320$ min. Since there are 390 min per day, the seventh scale corresponds to approximately one day.

it is not necessary to assume a particular distribution of the error term as in the GARCH/SV model.

As can be seen in Fig. 2.3, the hedge ratio at second scale has the lowest value, -0.06 and increases monotonically at a decreasing rate, converging toward the long horizon hedge ratio of one. As indicated in Fig. 2.3, the degree of hedging effectiveness has the lowest value at the second scale, 0.004, and approaches one as the wavelet time scale increases. Intuitively, hedging effectiveness approaches one because, over long horizons, the shared permanent component ties the stock and futures series together, and the effect of the transitory components becomes negligible. In the long run, the stock and futures prices are perfectly correlated (Geppert, 1995). This result is consistent with the results of Low *et al.* (2002) and Geppert (1995), who compare the hedge ratios and hedging effectiveness obtained from various models.

2.4. Concluding Remarks

This chapter investigates the multiscale relationship between the stock and futures markets over various time horizons. Ours is the first analysis to undertake this investigation using wavelet analysis and high frequency intra-day data. The chapter examines this relationship in three ways: (1) the lead-lag relationship, (2) covariance/correlation, and (3) the hedge ratio and hedging effectiveness. To examine the lead-lag relationship between the two markets, we employ the Granger causality test for various time scales utilizing the decomposed data. The wavelet correlation is estimated by testing the correlation between the two markets in the various time scales from the wavelet coefficients. The hedge ratio, defined by the covariance between the stock return and the futures return divided by the volatility of the futures return, is calculated from the wavelet covariance and variance. The main advantage of using wavelet analysis is the ability to decompose the data into various time scales. This advantage allows researchers to investigate the relationship between two variables in various time scales, while the traditional methodology allows examination of only two time scales: short- and long-run scales.

The wavelet analysis is undertaken using the LA(8) wavelet filter, and it supports four main conclusions. First, it is found that at 5-minute (original data) and approximately one-day dynamics (seventh scale), the futures market Granger causes the stock market. From this result, it is concluded that the futures market is more efficient. In other words, price discovery is greater in the futures market.

Second, it is found that there is an approximate decreasing linear relationship between the wavelet variance and the wavelet scale, indicating the potential for long memory in the volatility series. This result implies that an investor with a short investment horizon has to respond to every fluctuation in the realized returns, while for an investor with a much longer horizon, the long-run risk is significantly less.

Third, at the eighth scale (associated with $128 \times 5 = 640$-minute changes), it is found that there is an apparent break in the variance for both series. Since there are 390 minutes in one trading day, the eighth scale corresponds to 1.64 days. This implies that the stock and futures returns have a multiscaling behavior. This apparent multiscaling behavior occurs approximately at the daily horizon, where the intra-day seasonality is strongly present.

Fourth, we also find that at the intra-day scales (up to seventh scale), the correlation coefficients between the two markets vary at different horizons, while at one-day and higher scales, the correlation remains very high and stabilizes close to one. In addition, the magnitude of the wavelet correlation between the two markets increases as the time scale increases, indicating that the stock and futures markets are not fundamentally different.

Fifth, the analysis of the wavelet hedge ratio and hedging effectiveness indicates that the hedge ratio at the second scale has the lowest value and increases monotonically at a decreasing rate, converging toward the long horizon hedge ratio of one. The degree of hedging effectiveness also has the lowest value at the second scale and approaches one as the wavelet time scale increases. Intuitively, hedging effectiveness approaches one because, over long horizons, the shared permanent component ties the stock and futures series together and the effect of the transitory components becomes negligible.

Chapter 3

Modeling the International Links Between the Dollar, Euro and Yen Interest Rate Swap Markets Through a Multiscaling Approach

This chapter investigates the links between the world's largest interest rate swap markets, namely the US dollar, the euro and the Japanese yen, over various time horizons through a wavelet multiscaling approach. In contrast to approaches used in previous studies, wavelet analysis allows us to decompose the data into various time scales. First, we investigate the issue of causal relationships across the three major international swap markets using a multiscaling approach. Using this technique, we find that the dynamic causal interactions intensified over time and are persistent after the d3 wavelet time scale, which corresponds to 8–16 trading days. We find that the variances of the dollar and the euro swap markets are high compared to the yen market. We also find that the correlation between swap markets varies over time but remains very high, especially between the dollar and the euro. However, it is much lower between the euro and the yen, and between the dollar and the yen, implying that even though there have been striking developments in international swap markets since 1999, the yen market remains relatively less integrated with the other major swap markets. Finally, there is a noticeable variability in the euro swap market, compared to the dollar and yen swap markets, regardless of the time scale.

3.1. Introduction

One of the largest components of the global derivatives markets and a natural adjunct to the fixed income markets is the interest rate swaps market. Twenty years ago, interest rate swaps were an obscure and specialized transaction. Today, they are one of the most heavily traded financial contracts. In 1982, the notional amount of swaps outstanding was almost zero; ten years later, it had reached $3 trillion; in June 2003, it reached $95 trillion. The swap market is now so important that some commentators have suggested it could take over the fundamental pricing role of the government bond market. Despite the obvious significance of swaps, the empirical research literature is at present very small compared to the literature on equity and conventional debt markets.

Most empirical studies of interest rate swaps have examined US dollar swaps. These studies include Brown, Harlow and Smith (1994); Collin-Dufresne and Solnik (2001); Duffie and Huang (1996); Duffie and Singleton (1997); Li and Mao (2003); In, Brown and Fang (2003b); Lang, Litzenberger and Liu (1998); and Sun, Sundaresan and Wang (1993). Despite the huge size of the global swap market, there is relatively little evidence on the pricing of interest rate swaps other than US dollar swaps. We know of only four published studies on other currencies: In, Brown and Fang (2004) on the Australian and US dollar; Brown, In and Fang (2002) on the Australian dollar; Eom, Subrahmanyam and Uno (2002) on the Japanese yen; and Lekkos and Milas (2001) on the British pound.

Recently, there has been a little-noted but potentially highly significant change in the structure of the world's swap market. In an historic shift, the size of the euro swap market now comfortably exceeds that of its US dollar counterpart.[1] Statistics available from the website of the Bank for International Settlements show that as at June 2003, the notional amount of euro interest rate swaps outstanding stood at the equivalent of $40.7 trillion, while the corresponding figure for US dollar swaps was $27.6 trillion. The notional amount of yen interest rate swaps outstanding was $13.5 trillion. Creation and expansion of the euro swap market have the potential to reshape the channels of influence between swap markets, rendering obsolete our current knowledge of the links between these markets.

During the period in which the swap markets developed rapidly, international capital markets were becoming increasingly globalized. In this environment, economic and financial commentators frequently assume that changes in one country's interest rate market will directly and rapidly affect those in another. Although there are many studies of short-term links between international stock markets, much less has been written on short-term links between debt markets. In part, this lack of evidence is due to the focus of the interest rate literature on real rather than nominal interest rates. A partial exception is Al Awad and Goodwin (1998), whose

---

[1]Trichet mentions but does not discuss "the huge expansion of the euro interest-rate swap market" (Trichet, 2001:8).

study includes analysis of weekly changes in nominal interest rates.[2] Using impulse response analysis, they conclude that shocks in nominal US interest rates evoke significant responses in the UK, France and Germany but not *vice versa*.

We have chosen to study the swap market because, unlike many other financial contracts, the interest rate swap contract is both internationally standardized and heavily traded. It is therefore an ideal vehicle to study the international links between interest rate markets. There are only three published studies of the short-run international links between interest rate swap markets, none of which includes the euro: Eom, Subrahmanyam and Uno (2002) using weekly data on the US and Japanese markets; In, Brown and Fang (2003a) using daily data on the US, UK and Japanese markets; and Lekkos and Milas (2001) using weekly data on the US and UK markets. It is an intriguing area of study because it provides insights into information transmission, and the pricing of swaps with different maturities.

The main purpose of this chapter is to deepen our knowledge of the links between the three major international swap markets: the US dollar, the euro and the Japanese yen. Swap market participants that belong to a diverse group includes intraday traders, hedge funds, portfolio managers, commercial banks, large corporations and central banks. These participants operate on very different time scales. Owing to the different decision-making time scales among traders, the true dynamic and causal relationships between international swap markets vary over different time scales associated with these different horizons. As a result, low-frequency and high-frequency shocks will affect the market differently. Thus, we need an econometric method that can capture the underlying dynamic structure at different time scales, which in turn involves the separation of local dynamics from global dynamics such as the market's long-term growth path, and transitory changes from permanent changes. Traditional methods are poorly equipped for this task because they use only two time scales (the long run and the short run). This chapter contributes to the literature on the causal relationships by using a wavelet mutliscaling method that decomposes a given time series on a scale-by-scale basis. The wavelet covariance decomposes the covariance between two stochastic processes over different time scales. A wavelet covariance in a particular

---

[2]Most of their analysis concerns real interest rates, calculated using weekly inflation rates that are interpolated from monthly data.

time scale indicates the contribution to the covariance between two stochastic variables. This feature of wavelet analysis allows us to examine the covariance/correlation over different time scales.

The remainder of the chapter is organized as follows. Section 3.2 discusses the data and provides descriptive statistics. Section 3.3 presents the empirical results and discusses the findings. Finally, Section 3.4 provides a summary and conclusion.

3.2. Data and Descriptive Statistics

We use daily closing mid-rate data on swap maturities of three, five and ten years, for the US dollar, the euro and the Japanese yen in the period January 4, 1999 to January 31, 2003, giving a sample size of 1065 observations for each swap maturity. Data for each currency were collected from Datastream.

Descriptive statistics are reported in Panel A of Table 3.1. On average, the yield curves for swap rates are upward sloping in the three currencies during the sample period. The variances range from 0.832 to 2.133 for dollar swap rates, from 0.072 to 0.185 for yen swap rates and from 0.262 to 0.502 for euro swap rates. The measures for skewness and kurtosis are also reported to indicate whether swap rates are normally distributed. The sign of skewness varies between countries and maturities. In all cases, the Jarque-Bera statistic (denoted by JB) rejects normality at any conventional level of statistical significance. The Ljung-Box statistic for 15 lags [denoted by LB(15)] and squared term [denoted by $LB^2(15)$] indicate that significant linear and nonlinear dependencies exist.

Panel B of Table 1 presents cross-correlations for swap rates for the three different maturities in the three currencies. The correlation structure of these swap markets is perhaps the most important feature from the viewpoint of investors and portfolio managers because hedging and diversification strategies invariably involve some measure of correlation (In and Kim, 2006). Overall, the correlations between markets vary with maturity. The strongest correlation is observed between the dollar and the yen, regardless of the maturity, followed by the dollar and the euro, and finally the euro and the yen. The correlations range from 0.827 to 0.865 for the dollar–yen, from 0.600 to 0.729 for the dollar–euro, and from 0.366 to 0.400 for the yen–euro.

Table 3.1. Preliminary statistics.

Panel A. Descriptive statistics

	US Dollar			Euro			Japanese Yen		
	3-year	5-year	10-year	3-year	5-year	10-year	3-year	5-year	10-year
Mean	5.311	5.683	6.127	4.365	4.701	5.230	0.485	0.860	1.674
Variance	2.133	1.445	0.832	0.502	0.402	0.262	0.072	0.155	0.185
Skewness	−0.281	−0.297	−0.195	−0.220	−0.408	−0.540	0.247	0.077	−0.105
Kurtosis	−1.019	−0.815	−0.785	−0.845	−0.734	−0.649	−1.561	−1.622	−1.281
JB	60.103	45.124	34.113	40.313	53.445	70.475	118.886	117.786	74.769
	(0.000)	(0.000)	(0.000)	(0.000)	(0.000)	(0.000)	(0.000)	(0.000)	(0.000)
LB(15)	15424.404	20080.098	19659.073	19277.968	18942.066	18511.819	19300.468	19797.507	19385.871
	(0.000)	(0.000)	(0.000)	(0.000)	(0.000)	(0.000)	(0.000)	(0.000)	(0.000)
LB^2(15)	20532.641	20227.587	19752.334	19471.121	19098.794	18627.753	18062.089	19173.404	19310.049
	(0.000)	(0.000)	(0.000)	(0.000)	(0.000)	(0.000)	(0.000)	(0.000)	(0.000)

Note: The statistics refer to swap rates (% pa) using daily closing mid-rate data for the period January 4, 1999 to January 31, 2003 for US dollar, euro and Japanese yen. Data were collected from Datastream. Significance levels are in parentheses. LB(n) is the Ljung–Box statistic for up to n lags, distributed as χ^2 with n degrees of freedom. Skewness and kurtosis are defined as $E[(R_t - \mu)]^3$ and $E[(R_t - \mu)]^4$, where μ is the sample mean. JB indicates the Jarque–Bera statistic.

Table 3.1. (*Continued*)

Panel B. Cross-correlations between swap rates

	Dollar	Dollar $(t-1)$	Euro	Euro $(t-1)$	Yen	Yen $(t-1)$
3-year swap rates						
Dollar	1.000	0.999	0.600	0.594	0.858	0.856
Dollar $(t-1)$		1.000	0.602	0.598	0.858	0.856
Euro			1.000	0.998	0.406	0.404
Euro $(t-1)$				1.000	0.402	0.400
Yen					1.000	0.997
Yen $(t-1)$						1.000
5-year swap rates						
Dollar	1.000	0.998	0.638	0.631	0.865	0.864
Dollar $(t-1)$		1.000	0.640	0.636	0.865	0.863
Euro			1.000	0.997	0.367	0.367
Euro $(t-1)$				1.000	0.362	0.362
Yen					1.000	0.997
Yen $(t-1)$						1.000
10-year swap rates						
Dollar	1.000	0.997	0.729	0.722	0.827	0.825
Dollar $(t-1)$		1.000	0.731	0.728	0.825	0.824
Euro			1.000	0.997	0.366	0.365
Euro $(t-1)$				1.000	0.359	0.359
Yen					1.000	0.997
Yen $(t-1)$						1.000

Note: The correlations are calculated using daily closing mid-rate swap rates (% pa) data for the period January 4, 1999 to January 31, 2003 for US dollar, euro and Japanese yen. Data were collected from Datastream.

3.3. Empirical Results

In order to test for Granger-causality between the three swap markets at different time scales, a MODWT multiresolution analysis was performed using the Daubechies LA(8) wavelet filter. Since we use daily data, the wavelet scales d1, d2, d3, d4, d5, d6, and d7 are associated with oscillation of periods 2–4, 4–8, 8–16, 16–32, 32–64, 64–128 and 128–256 days, respectively.

Tables 3.2 and 3.3 report the Granger-causality test of swap rates and swap volatilities, respectively. Both tests are based on the wavelet domain for the three different maturities. The results reported in these tables show that there are causality effects between pairs of swap markets even in the first layer, which represents a period of 2 days. Overall, however, the causality effects show strongly beyond the third layer, which represents a data length of around 8 days. It follows that the three swap markets become more strongly related as the time-scale increases. For example, Table 3.2

Table 3.2.   Granger-causality test of swap rates on the wavelet domain.

Panel A. 3-year swap rates

	d1	d2	d3	d4	d5	d6	d7
EUR → USD	0.034	3.903*	4.839*	4.000*	3.383*	20.987*	1.599
	(0.853)	(0.020)	(0.028)	(0.046)	(0.034)	(0.000)	(0.206)
USD → EUR	1.296	0.756	4.442*	1.993	6.924*	21.339*	1.793
	(0.255)	(0.470)	(0.035)	(0.158)	(0.001)	(0.000)	(0.181)
JPY → USD	1.505	0.001	0.786	8.594*	3.308*	0.429	95.165*
	(0.220)	(0.973)	(0.502)	(0.003)	(0.037)	(0.513)	(0.000)
USD → JPY	1.859	0.852	6.600*	8.594*	1.954	0.375	84.400*
	(0.173)	(0.356)	(0.000)	(0.003)	(0.142)	(0.540)	(0.000)
JPY → EUR	0.001	1.166	4.894*	0.011	1.402	2.893*	8.424*
	(0.975)	(0.322)	(0.027)	(0.917)	(0.201)	(0.003)	(0.004)
EUR → JPY	1.220	2.565*	4.084*	0.002	6.265*	4.836*	6.530*
	(0.270)	(0.018)	(0.044)	(0.962)	(0.000)	(0.000)	(0.011)

Note: The significance levels are in parentheses and * indicates significance at 5% level.

Panel B. 5-year swap rates

	d1	d2	d3	d4	d5	d6	d7
EUR → USD	0.360	2.378	0.851	1.796	0.549	35.913*	11.713*
	(0.549)	(0.093)	(0.493)	(0.097)	(0.459)	(0.000)	(0.001)
USD → EUR	4.552*	2.805	1.262	2.150*	0.446	36.005*	12.561*
	(0.033)	(0.061)	(0.283)	(0.046)	(0.504)	(0.000)	(0.000)
JPY → USD	2.823	1.696	16.537*	4.440*	8.667*	3.951*	127.260*
	(0.093)	(0.184)	(0.000)	(0.035)	(0.000)	(0.047)	(0.000)
USD → JPY	0.001	5.116*	18.949*	5.686*	3.816*	3.108	105.212*
	(0.976)	(0.006)	(0.000)	(0.017)	(0.022)	(0.078)	(0.000)
JPY → EUR	0.739	1.707	0.010	0.006	1.959	0.409	25.319*
	(0.390)	(0.116)	(0.921)	(0.939)	(0.058)	(0.523)	(0.000)
EUR → JPY	0.634	2.467*	0.001	0.101	3.914*	0.113	20.052*
	(0.426)	(0.022)	(0.980)	(0.750)	(0.000)	(0.737)	(0.000)

Note: The significance levels are in parentheses and * indicates significance at 5% level.

Panel C. 10-year swap rates

	d1	d2	d3	d4	d5	d6	d7
EUR → USD	1.057	3.758*	0.174	9.375*	7.623*	65.069*	99.982*
	(0.304)	(0.024)	(0.677)	(0.002)	(0.006)	(0.000)	(0.000)
USD → EUR	5.223*	3.064*	0.000	8.562*	7.644*	66.463*	102.677*
	(0.022)	(0.047)	(0.995)	(0.004)	(0.006)	(0.000)	(0.000)
JPY → USD	0.689	2.808	1.877	8.508*	3.649	9.220*	184.412*
	(0.407)	(0.094)	(0.171)	(0.004)	(0.056)	(0.002)	(0.000)
USD → JPY	0.088	3.189	0.323	6.119*	3.759	8.421*	146.127*
	(0.767)	(0.074)	(0.570)	(0.014)	(0.053)	(0.004)	(0.000)

(*Continued*)

Table 3.2.   (*Continued*)

	d1	d2	d3	d4	d5	d6	d7
JPY → EUR	1.022	2.811	1.687	14.661*	7.437*	2.226	42.680*
	(0.312)	(0.061)	(0.186)	(0.000)	(0.006)	(0.108)	(0.000)
EUR → JPY	0.478	2.198	6.506*	10.321*	8.744*	1.815	35.867*
	(0.489)	(0.112)	(0.002)	(0.001)	(0.003)	(0.163)	(0.000)

Note: The significance levels are in parentheses and * indicates significance at 5% level. Data used are daily closing mid-rate swap rates (% pa) for the period January 4, 1999 to January 31, 2003 for the US dollar, the euro and the Japanese yen. Data were collected from Datastream. Column headings (d1 to d7) refer to wavelet scales for oscillations of periods 2–4, 4–8, 8–16, 16–32, 32–64, 64–128 and 128–256 days. For example, the original data has been transformed by the wavelet filter [LA(8)] up to time scale 7 and the first detail (wavelet coefficient) d1 captures oscillation with a period length of 2 to 4 days. The last detail d7 captures oscillation with a period length of 128 to 256 days. EUR → USD indicates the Granger-causality test in the wavelet domain for the contemporaneous spillover effects of daily swap rates from the euro to the US dollar, USD → EUR from the US dollar to the euro, JPY → USD from the Japanese yen to the US dollar, USD → JPY from the US dollar to the Japanese yen, JPY → EUR from the Japanese yen to the euro, and EUR → JPY from the euro to the Japanese yen, respectively.

Table 3.3.   Granger-causality test of swap volatilities on the wavelet domain.

Panel A. 3-year swap volatilities

	d1	d2	d3	d4	d5	d6	d7
EUR → USD	0.004	1.234	4.903*	11.097*	7.899*	5.942*	1.521
	(0.952)	(0.267)	(0.000)	(0.001)	(0.000)	(0.015)	(0.145)
USD → EUR	0.285	0.087	3.113*	10.315*	0.898	8.010*	6.293*
	(0.594)	(0.768)	(0.009)	(0.001)	(0.408)	(0.005)	(0.000)
JPY → USD	1.269	1.194	43.546*	5.891*	24.091*	1.360	188.050*
	(0.260)	(0.303)	(0.000)	(0.015)	(0.000)	(0.254)	(0.000)
USD → JPY	1.340	2.357	28.071*	3.343	22.547*	3.536*	94.282*
	(0.247)	(0.095)	(0.000)	(0.068)	(0.000)	(0.014)	(0.000)
JPY → EUR	0.604	0.073	0.902	0.592	26.566*	15.847*	3.545*
	(0.437)	(0.788)	(0.462)	(0.737)	(0.000)	(0.000)	(0.014)
EUR → JPY	1.723	0.271	1.615	2.057	23.495*	21.234*	3.584*
	(0.190)	(0.603)	(0.168)	(0.056)	(0.000)	(0.000)	(0.013)

Note: The significance levels are in parentheses and * indicates significance at 5% level.

Panel B. 5-year swap volatilities

	d1	d2	d3	d4	d5	d6	d7
EUR → USD	0.049	0.750	0.142	10.061*	17.686*	1.031	2.397
	(0.825)	(0.387)	(0.706)	(0.002)	(0.000)	(0.378)	(0.091)

(*Continued*)

Table 3.3. (*Continued*)

	d1	d2	d3	d4	d5	d6	d7
USD → EUR	0.197	0.290	0.004	6.032*	13.484*	8.937*	18.577*
	(0.657)	(0.591)	(0.947)	(0.014)	(0.000)	(0.000)	(0.000)
JPY → USD	0.287	4.398*	34.615*	2.427	71.497*	7.749*	6.789*
	(0.751)	(0.013)	(0.000)	(0.089)	(0.000)	(0.000)	(0.000)
USD → JPY	1.084	2.018	23.584*	8.736*	56.131*	4.084*	4.846*
	(0.339)	(0.133)	(0.000)	(0.000)	(0.000)	(0.000)	(0.002)
JPY → EUR	4.544*	2.916*	2.548*	34.443*	107.460*	96.990*	9.540*
	(0.033)	(0.033)	(0.027)	(0.000)	(0.000)	(0.000)	(0.002)
EUR → JPY	6.356*	3.627*	2.603*	44.723*	91.437*	140.045*	5.303*
	(0.012)	(0.013)	(0.024)	(0.000)	(0.000)	(0.000)	(0.021)

Note: The significance levels are in parentheses and * indicates significance at 5% level.

Panel C. 10-year swap volatilities

	d1	d2	d3	d4	d5	d6	d7
EUR → USD	0.483	1.975	0.000	35.581*	12.077*	0.421	162.778*
	(0.487)	(0.160)	(0.997)	(0.000)	(0.001)	(0.517)	(0.000)
USD → EUR	0.318	3.205	0.204	26.783*	8.785*	0.213	131.133*
	(0.573)	(0.074)	(0.652)	(0.000)	(0.003)	(0.644)	(0.000)
JPY → USD	0.591	1.105	3.915*	3.488*	6.361*	11.475*	6.209*
	(0.554)	(0.293)	(0.004)	(0.001)	(0.012)	(0.000)	(0.000)
USD → JPY	1.453	4.748*	1.068	5.128*	5.273*	9.892*	3.149*
	(0.234)	(0.030)	(0.371)	(0.000)	(0.022)	(0.000)	(0.002)
JPY → EUR	1.223	1.138	1.329	1.464	7.440*	235.625*	24.149*
	(0.296)	(0.333)	(0.257)	(0.176)	(0.006)	(0.000)	(0.000)
EUR → JPY	2.513*	2.940*	1.427	0.729	6.999*	275.184*	18.620*
	(0.028)	(0.032)	(0.223)	(0.647)	(0.008)	(0.000)	(0.000)

Note: The significance levels are in parentheses and * indicates significance at 5% level. Data used are daily closing mid-rate swap rate volatilities (% pa) for the period January 4, 1999 to January 31, 2003 for the US dollar, the euro and the Japanese yen. Data were collected from Datastream. Column headings (d1 to d7) refer to wavelet scales for oscillations of periods 2–4, 4–8, 8–16, 16–32, 32–64, 64–128 and 128–256 days. For example, the original data has been transformed by the wavelet filter [LA(8)] up to time scale 7 and the first detail (wavelet coefficient) d1 captures oscillation with a period length of 2 to 4 days. The last detail d7 captures oscillation with a period length of 128 to 256 days. EUR → USD indicates the Granger-causality test in the wavelet domain for the contemporaneous spillover effects of daily swap rate volatilities from the euro to the US dollar, USD → EUR from the US dollar to the euro, JPY → USD from the Japanese yen to the US dollar, USD → JPY from the US dollar to the Japanese yen, JPY → EUR from the Japanese yen to the euro, and EUR → JPY from the euro to the Japanese yen, respectively. Volatility is measured as the absolute value of the first difference of each swap rate.

shows that the Granger-causality of 3-year swap rates is strong, evidenced by the fact that four out of the six connections at the longest (d7) wavelet scale are statistically significant, while all six connections for both 5-year swap rates and 10-year swap rates are statistically significant. Similarly, the results in Table 3.3 imply that the Granger-causality of swap volatilities is strongest at the longest (d7) wavelet scale, evidenced by the statistical significance of five out of the six connections for both 3-year and 5-year swap volatilities, and all six connections for 10-year swap volatilities. In short, the key common finding from Tables 3.2 and 3.3 is that, overall, the wavelet multiscale Granger-causality tests indicate that the connections among all three swap markets are stronger and more certain in the long term, regardless of swap maturity.

We analyze the relationships between pairs of swap markets using wavelet analysis. To be consistent with the Granger-causality tests, we use the MODWT based on the Daubechies least asymmetric family of wavelets [LA(8)]. We investigate the swap market links in terms of wavelet covariances and correlations over the various time scales. To investigate the multiscale correlation, we report the variances of swap market rates for the three currencies, for the three different swap maturities.

Fig. 3.1 illustrates the MODWT-based wavelet variance of dollar swap rates, euro swap rates and yen swap rates. The wavelet variances show that the 3-year swap market is more volatile than the 5-year and 10-year swap markets regardless of the time scale. This pattern is consistent across the different currencies, supporting the conventional perception that shorter-term rates tend to be more volatile than longer-term rates.

In addition to examining the variances of currencies, a natural question is how different currency series are associated with one another. Note that a wavelet covariance in a particular time scale indicates the contribution to the covariance between two series. Fig. 3.2 shows the MODWT-based wavelet covariances for 3-year, 5-year and 10-year swap rates and for pairs of swap markets (i.e., dollar *vs* euro, dollar *vs* yen, and euro *vs* yen) using the LA(8) wavelet filter.

Overall, the movements of covariance decrease as the time scale increases. However, it is difficult to compare wavelet covariances between maturities or between currencies because of the different variability exhibited by them. Therefore, we examine the wavelet correlations.

Fig. 3.3 shows the correlation between pairs of swap markets (i.e., dollar *vs* euro, dollar *vs* yen, and euro *vs* yen) against the wavelet scales for 3-year, 5-year and 10-year swap rates. A distinctive feature of Fig. 3.3 is the high

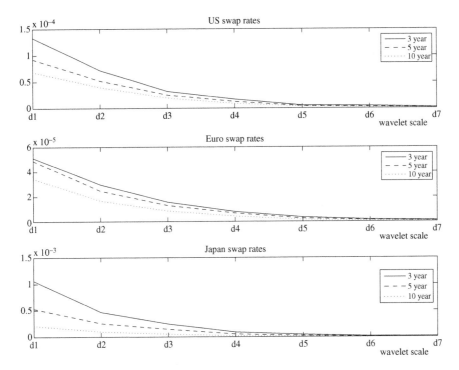

Fig. 3.1. Wavelet variances of US dollar, euro and Japanese yen swap rates.
Note: Estimated wavelet variances of US dollar, euro and Japanese yen swap rates plotted
on different wavelet time-scales on the horizontal axis. The vertical axis indicates the
wavelet variance. The MODWT-based wavelet variances of US dollar, euro and Japanese
yen swap rates have been constructed using the Daubechies least asymmetric wavelet
filter of length 8 [LA(8)].

and significantly positive relationship that can be observed in all time scales
in the case of the dollar *vs* the euro. The correlation between these markets
steadily increases over time and, at more than 0.6 on average, remains very
high.

In contrast, the correlations between the dollar and the yen, and
between the euro and the yen, show a low and (usually) positive relationship
in all time scales. For the 3-year and 5-year swap rates, the correlations
between the dollar and the yen, and between the euro and the yen increase
slightly over time, up to the 4th layer, which represents a data period of
around 16 days. After this point, the series falls into low and negative
relationships. Similarly, for 10-year swap rates, the correlations between
the dollar and the yen, and between the euro and the yen, remain very

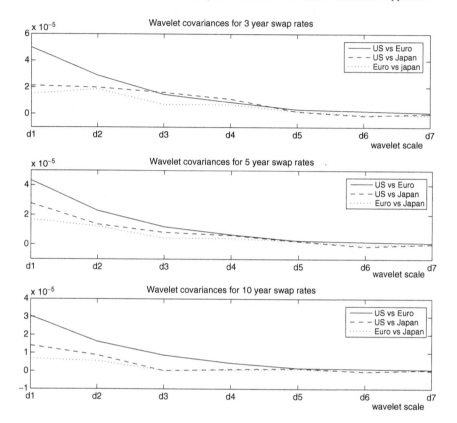

Fig. 3.2.   Wavelet covariances of US dollar, euro and Japanese yen swap rates.
Note: Estimated wavelet covariances of US dollar, euro and Japanese yen swap rates
plotted on different wavelet time-scales on the horizontal axis. The vertical axis indicates
the wavelet covariance. The MODWT-based wavelet covariances of US dollar, euro and
Japanese yen swap rates have been constructed using the Daubechies least asymmetric
wavelet filter of length 8 [LA(8)].

low and positive until the 5th layer, turning negative at the 6th layer. It is
positive at the 7th layer for the dollar-yen pair but remains negative for
the euro–yen pair. The low magnitude of the correlations involving the yen
imply that the yen swap market is less tightly connected with the other
major swap markets.

  Figs. 3.4(a), 3.4(b) and 3.4(c) show the LA(8) MODWT MRA of 10-
year swap rates for the dollar, the euro and the yen respectively, using
various time scales (i.e., the seven different wavelet details, d1 to d7). In
each plot, the topmost panel shows the original time series. Below it, from

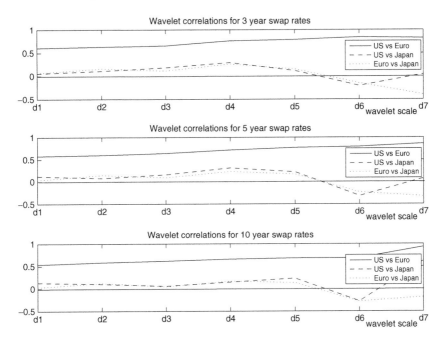

Fig. 3.3. Wavelet correlations of US dollar, euro and Japanese yen swap rates.
Note: The MODWT-based wavelet correlations have been constructed using the Daubechies LA(8) wavelet filter. Overall, the estimated wavelet correlation between US dollar and euro swap rates shows a steadily increasing pattern throughout the scales d1 to d7, regardless of the swap term. In contrast, estimated wavelet correlations between US dollars and yen, and between euros and yen, increase very slowly until the d4 scale (peaking at a correlation of about 0.2), then plunge down until around the d6 scale, beyond which it increases or decreases depending on the currency pair and the term.

top to bottom, are the wavelet details d1, d2,..., d7. One of the distinctive features of the MRA of the three swap markets is that there are many peaks in the original series for the euro swap market, compared to the dollar and yen swap markets. This feature is captured in the d1 and d2 components. At the highest time scale, d7, representing the deviation from the long-term trend, overall there is relatively smooth movement in each currency. However, the long-term movement of the euro swap market is different from that of the dollar and yen swap markets. The deviation from the long-term movement of the euro swap market has more pronounced peaks and troughs, reflecting the less stable nature of the market, even in the long-term scales. For those market participants who are operating in the international swap markets on longer time horizons, such as central banks, this empirical finding may lead to better informed views on the behavior of swap rates.

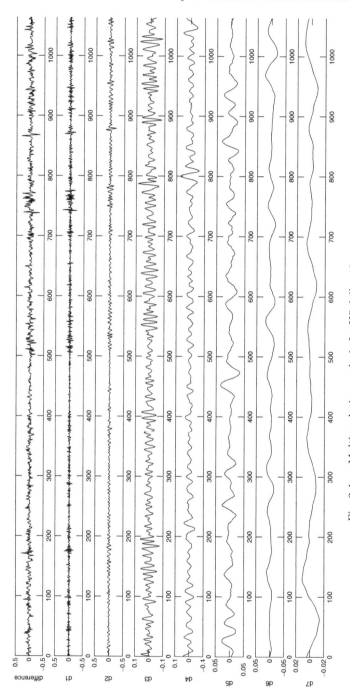

Fig. 3.4a.   Multiresolution analysis for US dollar 10-year swap rate.

Note: LA (8) MODWT MRA of the US dollar 10-year swap rates. In each plot, the upper row is the original time series. Below it — from top to bottom — are the wavelet details d1,...,d8.

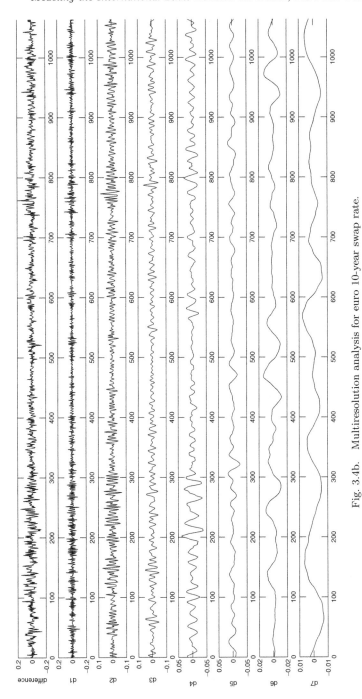

Fig. 3.4b. Multiresolution analysis for euro 10-year swap rate.

Note: LA (8) MODWT MRA of the euro 10-year swap rates. In each plot, the upper row is the original time series. Below it — from top to bottom — are the wavelet details d1,..., d8.

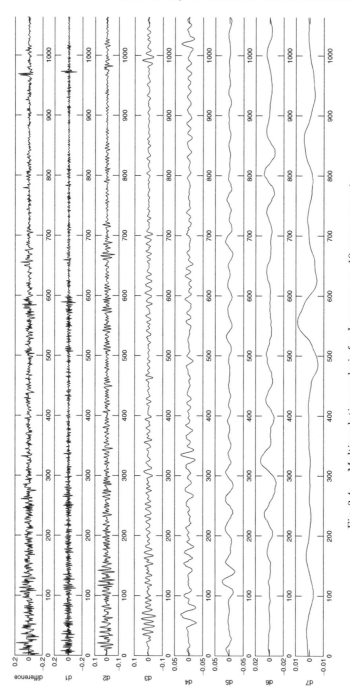

Fig. 3.4c. Multiresolution analysis for Japanese yen 10-year swap rate.

Note: LA (8) MODWT MRA of the Japanese yen 10-year swap rates. In each plot, the upper row is the original time series. Below it — from top to bottom — are the wavelet details d1, . . . , d8.

Overall, as the time scale increases from the finest time scale (d1) to the coarsest time scale (d7), the wavelet coefficients show a progressively smoother movement, implying that short-term noise in the market is eliminated and consequently there emerges the "true" underlying economic relationship between swap markets. These wavelet MRA figures indicate that the wavelet decomposition provides information that cannot be captured by conventional analysis. In other words, the decomposition of data into several time scales is important in economics and finance, since it detects the frequency burst in various time scales.

3.4. Concluding Remarks

This chapter examines the links between the major international interest rate swap markets, namely the US dollar, the euro and the Japanese yen over various time horizons. We propose a new approach — the wavelet multiscaling method — to undertake this investigation. The approach focuses on the relationship in four ways: (1) the lead-lag causal relationship; (2) variance (covariance); (3) correlations; and (4) multiresolution analysis.

To examine the lead-lag relationship between the two markets, we employ the Granger-causality test for various time scales across the three major international swap markets using a wavelet multiscaling method. Using this technique, we find that the dynamic causal interactions intensify over time. It is found that beyond the third-level time scale, the three major swap markets show more active feedback relationships regardless of swap maturity.

Second, it is found that there is an approximate linear relationship between the wavelet variance and the wavelet scale, indicating the potential for long memory in the swap volatility series. Wavelet variances and covariances decrease as the wavelet time scale increases. Overall, the wavelet variances show that the dollar and the euro swap markets are more volatile than their yen counterparts, regardless of the time scale.

Third, we also find that the correlation between the swap markets varies over time but remains very high, especially for the euro and dollar swap markets but is much lower between the euro and yen swap markets, and between the dollar and yen swap markets. This finding implies that the yen swap market is relatively less integrated with the other major swap markets.

Fourth, according to the MODWT multiresolution analysis, we also find that there is a noticeable variability in the euro swap market, compared to the dollar and yen swap markets, regardless of the different time scales. The analysis shows that introduction of the euro swap market provides an insight for central banks, in particular, which operate on larger time horizons in the international swap markets.

Chapter 4

Long Memory in Rates and Volatilities of LIBOR: Wavelet Analysis

This chapter examines the long-memory behavior of the LIBOR rates and volatilities using wavelet analysis. The major innovation of this chapter is to introduce the wavelet OLS estimation approach (Jensen, 1999a) using London Interbank Offer Rate (LIBOR) data for four major currencies, namely, the US dollar, the UK pound, the Japanese yen, and the Australian dollar. Our empirical findings show that the Japanese yen and Australian dollar LIBOR rates do not show any significant long-memory pattern, while the US dollar and UK pound do show some degree of significant long-memory pattern regardless of the maturity. In contrast, for all currencies, we find that the volatility has a clear pattern of long memory regardless of maturity and currencies, except for 6-month LIBOR rates for the Japanese yen and Australian dollar.

4.1. Introduction

Recently, several works have found evidence of stochastic long-memory behavior in stock returns and volatility. If there is long memory in return and volatility, it is possible to obtain increased profits on the basis of price change predictions that contradict the efficient market hypothesis. The presence of long-memory dynamics, which is a special form of nonlinear relationships, gives nonlinear dependence in the first moment of the distribution and, hence, a potentially predictable component in the series dynamics. Fractionally integrated processes can give rise to long-term memory. On the other hand, the short-memory property describes the low correlation structure of a time series. For short-memory series, correlations among observations at long lags become negligible (Barkoulas *et al.*, 1997). The empirical presence of long memory is found in the persistence of the autocorrelations. This slow decaying pattern of the autocorrelation is not consistent either with the stationary, short-memory, ARMA models, nor with the non-stationary, unit root models. As is pointed out in Jensen (1999a), the long-memory approach falls nicely in between these two knife-edge approaches.

For the tests of long memory in the financial markets, there are at least three common tests in the literature; (i) the R/S-Statistic (rescaled adjusted range), (ii) the nonparametric spectral test, and (iii) Geweke and Porter-Hudak (1983, hereafter GPH). First, the R/S-statistic, which arose from the pioneering work by Hurst (1951) and was further developed and modeled by Mandelbrot's work (1972), has been widely used to analyze the fractal behavior and stochastic memory of financial time series. Using this method, Booth *et al.* (1982a, 1982b), among others, conclude that some financial time series have long-memory behavior. However, Lo (1991) used a version of the R/S statistic of Hurst (1951) and showed that there was no evidence of long-range dependence of the monthly and daily returns on CRSP stock indices.

Second, nonparametric spectral analysis is a method used in the frequency domain, rather than the time domain. Lobato and Robinson (1998) examine the presence of long memory in the stock market using spectral analysis, and find no evidence for long memory in returns. Crato and Ray (2000) adopt spectral analysis to examine the existence of long memory in futures' return and volatility. They reach a similar conclusion to Lobato and Robinson (1998).

Finally, GPH is a test based on the estimator of the long-memory parameter, and its adoption can be seen in much recent literature. GPH utilizes a nonparametric approach, which regresses the log values of the periodogram on the log Fourier frequencies to estimate the fractional differencing parameter. Barkoulas *et al.* (1997) use this methodology to test long memory in commodity prices, and find that characterization of the low-frequency properties for commodity prices based on integer integration tests can lead to misleading inference, arguing that the GPH testing procedure can allow for a rich pattern of interactions between short-term and long-term dynamics and that it is well-suited to capture long-term memory dynamics in the sample series. However, this estimator possesses no satisfactory asymptotic properties. For example, Priestley (1992:425) points out the inconsistency of the periodogram as an estimator of the spectrum. In addition, Hurvich and Beltrao (1993) and Robinson (1995) suggest that the normalized periodogram is neither asymptotically independent nor identically distributed.

Recently, wavelet analysis has been adopted to examine the long-term dependence of time series. Jensen (1999a) proposes a new methodology to examine long memory using wavelets. Wavelets can be most simply described as functional transforms in the same spirit as Fourier transforms, but with properties that allow them to more effectively identify either long rhythmic behavior or short-run phenomena (Tkacz, 2001). In the study, he shows that the wavelet OLS (henceforth WOLS) estimator yields a consistent estimator of the fractional differencing parameter. Using this method, Tkacz (2001) and McCarthy *et al.* (2004) examine the long memory of interest rates.

In our study, we adopt Jensen's WOLS estimator to examine the long-memory behavior of four major currency LIBOR rates, namely, the US dollar, the UK pound, the Japanese yen and the Australian dollar, with four different maturities over various time horizons. This new approach is based on a wavelet multiscaling method that decomposes a given time series on a scale-by-scale basis. The main advantage of the wavelet analysis is the ability to simultaneously localize a process in time and scale.[1]

This chapter aims to investigate the long-memory behavior of LIBOR rates in four major currencies by estimating the fractional integration parameter, d, of the percentage change in selected LIBOR rates. In doing so, we adopt a methodology proposed by Jensen (1999a).

In this study, we first find that the LIBOR rates have very weak long-memory behavior, which is indicated by small values of the estimates for the long-memory parameters, d. In particular, the Australian dollar and Japanese yen do not show any significant long-memory parameter, while the US and UK data appear to display weak long-memory behavior. Using four different maturities over various horizons, it is also observed that the outcome of the test results does not depend much on maturities and sample periods. In the case of volatilities, all long-memory parameters of volatilities of LIBOR rates show the long-memory behavior, except 6-month LIBOR rates for the Australian dollar and Japanese yen.

---

[1]The key distinctive features of wavelet analysis are that wavelets possess not only the ability to perform nonparametric estimations of highly complex structures without knowing the underlying functional form, but also are able to accurately locate discontinuity and high frequency bursts in dynamic systems. In short, the major aspects of wavelet analysis are the ability to handle nonstationary data, localization in time, and the resolution of the signal in terms of the time scale of analysis. Among these aspects, the most important property of wavelet analysis is decomposition by time scale (Ramsey, 1999).

The organization of the chapter is as follows. In Section 4.2, we present the data and empirical results. A summary and concluding remarks are presented in Section 4.3.

4.2. Data and Empirical Results

Our data sets used in this empirical analysis consists of daily observations on BBA LIBOR rates. Large international banks actively trade with each other with 1-month, 3-month, 6-month, and 12-month deposits denominated in all of the world's major currencies. LIBOR is a widely used reference rate, and LIBOR rates are generally higher than the corresponding Treasury rates because they are not risk-free rates. These are the most popularly used data as proxies of short-term interest rates. In order to discuss international evidence on the long-memory behavior of interest rates, we use LIBOR rates in major currencies, namely the US dollar, the UK pound, the Japanese yen and the Australian dollar. We use daily LIBOR rates data on maturities of three, six, nine and twelve months, for the US dollar, the UK pound, the Japanese yen and the Australian dollar in the period March 20, 1995 to May 13, 2003, giving a sample size of 2049 observations[2] for each LIBOR rates maturity. Data for each currency were collected from Datastream. We compute the daily percentage change for each LIBOR rate calculated by $100 \times \{\log(r_t) - \log(r_{t-1})\}$.

Table 4.1 presents basic summary statistics for the interest rates level. A glance at the sample means of the interest rates indicates that all countries have an upward-sloping term structure. From the sample standard deviation, we can first observe that the Japanese yen LIBOR rates are more stable than the other currency LIBOR rates. It is worth noting that the short-term maturities have a lower variability than the long maturities. The signs of skewness are negative in the UK pound and the US dollar LIBOR rates, while positive in the Australian dollar and the Japanese yen LIBOR rates. In all cases, regardless of maturities, the Jarque–Bera statistic (denoted JB) rejects normality at any conventional level of statistical significance.

Before we discuss our main empirical results, we examine the autocorrelation for 3-month LIBOR rates up to 60 days in the four currencies.[3]

---

[2]Due to the restriction of DWT, we trim our sample to $2^{11} = 2048$ observations.
[3]We also plot the other maturities. However, the figure is not much different from those presented in the chapter. The other figures are available on request.

Table 4.1. Basic statistics.

		Mean	S.D	Skewness	Kurtosis	Minimum	Maximum	JB	JB-sig
US dollar	3 month	4.847	1.671	−1.108	−0.275	1.230	6.869	425.794	0.000
	6 month	4.898	1.685	−1.080	−0.293	1.170	7.109	405.602	0.000
	9 month	4.967	1.681	−1.055	−0.272	1.168	7.333	386.354	0.000
	12 month	5.058	1.665	−1.034	−0.234	1.190	7.501	369.690	0.000
UK pound	3 month	5.882	1.157	−0.370	−0.788	3.593	7.875	99.592	0.000
	6 month	5.941	1.169	−0.402	−0.792	3.518	7.938	108.557	0.000
	9 month	6.009	1.170	−0.437	−0.784	3.466	7.969	117.648	0.000
	12 month	6.089	1.172	−0.463	−0.762	3.443	8.000	122.830	0.000
Japanese yen	3 month	0.394	0.323	1.481	4.272	0.051	2.188	2307.171	0.000
	6 month	0.417	0.320	1.304	3.315	0.070	2.094	1519.118	0.000
	9 month	0.445	0.331	1.162	2.409	0.081	2.063	956.646	0.000
	12 month	0.469	0.346	1.063	1.697	0.086	2.063	631.963	0.000
Australian dollar	3 month	5.648	1.059	0.775	−0.699	4.120	8.313	246.611	0.000
	6 month	5.696	1.085	0.781	−0.558	3.969	8.750	234.920	0.000
	9 month	5.772	1.112	0.772	−0.486	3.932	9.063	223.568	0.000
	12 month	5.863	1.136	0.773	−0.399	3.947	9.375	217.654	0.000

Note: Significance levels are in parentheses. LB(n) is the Ljung-Box statistic for up to n lags, distributed as χ^2 with n degrees of freedom. Skewness and kurtosis are defined as $E[(R_t - \mu)]^3$ and $E[(R_t - \mu)]^4$, where μ is the sample mean.

As indicated before, if a time series has long memory, its autocorrelation decays very slowly to zero as the lag increases, reflecting the fact that the influence of the past values is quite persistent even for large lags. Fig. 4.1 presents the autocorrelation coefficients for 3-month LIBOR rates, while Fig. 4.2 shows the autocorrelation coefficients for volatilities[4] of 3-month LIBOR rates in four countries. From panel A of Fig. 4.1, we observe that the autocorrelation coefficients for 3-month LIBOR rates for the US dollar and UK pound show slowly decaying patterns, implying that the US dollar and UK pound LIBOR rates might have a long-memory process, in contrast to the Japanese yen and Australian dollar LIBOR rates. From panel B of Fig. 4.1, the 3-month LIBOR rates for the Australian dollar and Japanese yen show no clear pattern of slow decay as the time lag increases, indicating that these two LIBOR rates do not have a long-memory process.

Overall, the autocorrelation patterns vary with currency, but not with maturity, indicating that different currencies have revealed different behaviors of long memory in their LIBOR rates.

Panels A and B of Fig. 4.2 present the autocorrelation coefficients for volatilities of 3-month LIBOR rates in four currencies. In contrast, one of the salient features of autocorrelation coefficients highlights slowly decaying patterns for all currencies, implying that they might have a long-memory property.

Long-memory behaviors of LIBOR rates and volatility are further discussed by using the WOLS estimator. All computations are performed using the Matlab toolbox Wavekit of Ojanen (1998). As mentioned in Ball (1989), if the log difference has a stochastic memory, investors can increase their profit through prediction of price changes. However, most empirical studies show that financial returns on major markets do not show a long-term memory property (for example, see Crato and Ray, 2000; Cheung and Lai, 1993).

Applying the Daubechies wavelet filter of length 20 (D20)[5], the long-memory parameter, d, was estimated for each of the daily percentage change and volatility of the 3-, 6-, 9-, and 12-month LIBOR rates for the US dollar, the UK pound, the Japanese yen, and the Australian

---

[4]The daily volatility is measured by absolute value of the daily change of the LIBOR rates.
[5]Practically, the wavelet coefficients are calculated using the wavelet filters. According to Gençay *et al.* (2010), there are three aspects to be considered for a right wavelet filter: length of data, complexity of the spectral density function, and the underlying shape of features in the data. Based on these three aspects, we use the Daubechies wavelet filter of length 20.

Panel A. US dollar and UK pound

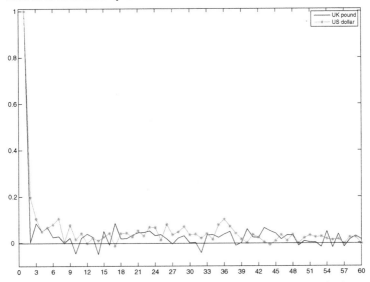

Note: The vertical axis indicates the autocorrelation coefficients, while the horizontal axis indicates time lags up to 60 days. The autocorrelation coefficients for 3-month LIBOR rates for the US dollar and the UK pound clearly show decaying patterns as the time lag increases.

Panel B. Japanese yen and Australian dollar

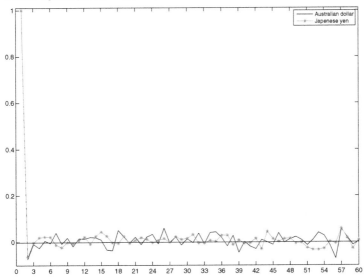

Note: The vertical axis indicates the autocorrelation coefficients, while the horizontal axis indicates time lags up to 60 days. Note that the autocorrelation coefficients for 3-month LIBOR rates for the Japanese yen and Australian dollar show no clear pattern of slow decay as the time lag increases.

Fig. 4.1. Autocorrelation coefficients of 3-month LIBOR rates.

Panel A. US dollar and UK pound

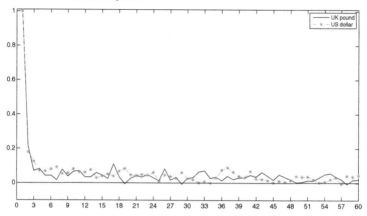

Note: The vertical axis indicates the autocorrelation coefficients, while the horizontal axis indicates time lags up to 60 days. The autocorrelation coefficients for 3-month LIBOR rates for the US dollar and the UK pound clearly show decaying patterns as the time lag increases.

Panel B. Japanese yen and Australian dollar

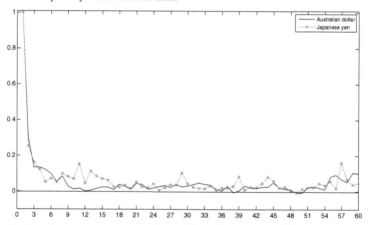

Note: The vertical axis indicates the autocorrelation coefficients, while the horizontal axis indicates time lags up to 60 days. Note that the autocorrelation coefficients for 3-month LIBOR rates for the Japanese yen and Australian dollar show no clear pattern of slow decay as the time lag increases.

Fig. 4.2.   Autocorrelation coefficients for volatilities of 3-month LIBOR rates.

dollar. Tables 4.2 and 4.3 report the long-memory estimates for LIBOR rates and volatilities of LIBOR rates, respectively. From Table 4.2, we observe that for the daily percentage change of LIBOR rates, two findings are worth noting. First, given the small values of the estimates for the long-memory parameters, d, the daily percentage changes appear to

Table 4.2.   Long-memory estimates for LIBOR rates.

		R-square	d	S.E.	t-value
US dollar	3 month	0.584	0.182*	0.051	3.557
	6 month	0.484	0.142*	0.049	2.905
	9 month	0.399	0.111*	0.045	2.443
	12 month	0.316	0.088*	0.043	2.039
UK pound	3 month	0.551	0.127*	0.038	3.321
	6 month	0.600	0.123*	0.034	3.674
	9 month	0.647	0.114*	0.028	4.059
	12 month	0.634	0.094*	0.024	3.945
Japanese yen	3 month	0.138	−0.059	0.049	−1.198
	6 month	0.024	−0.024	0.051	−0.474
	9 month	0.043	−0.033	0.053	−0.634
	12 month	0.048	−0.037	0.055	−0.675
Australian dollar	3 month	0.032	−0.030	0.054	−0.549
	6 month	0.120	−0.085	0.077	−1.109
	9 month	0.122	−0.119	0.106	−1.117
	12 month	0.166	−0.210	0.157	−1.338

Note: *indicates significance at 5% level. The results are estimated using the following equation.

$$\ln R(j) = \ln \sigma^2 - d \ln 2^{2j}$$

reveal weak long-memory behavior, although some of them are statistically significant. Second, the outcome of the result somewhat depends on the currencies under investigation. Specifically, the four currencies can be divided into two groups. The US dollar and UK pound belong to the first group, which shows evidence of statistically significant long-term dependence with 95% or better level of confidence. The second group consists of the Japanese yen and Australian dollar, which do not show long-memory behavior. For the US dollar, the results are similar to those of McCarthy *et al.* (2004), who examine the long-memory behavior for the full range of US government debt securities.

Table 4.3 reports the long-memory behavior of daily volatilities of LIBOR rates for the same four currencies. As in the case of the daily percentage change of LIBOR rates, the long-memory parameter of volatility is calculated using D20 wavelet filter. All long-memory parameters of volatilities of LIBOR rates show the long-memory behavior, except 6-month LIBOR rates for the Japanese yen and Australian dollar.

Clearly, our empirical results imply that considerable care should be taken in modeling the LIBOR rates. In fact, the previous literature seems

Table 4.3.   Long-memory estimates for volatilities of LIBOR rates.

		R-square	d	S.E.	t-value
US dollar	3 month	0.787	0.278*	0.048	5.762
	6 month	0.792	0.352*	0.060	5.861
	9 month	0.768	0.397*	0.073	5.461
	12 month	0.777	0.407*	0.073	5.603
UK pound	3 month	0.631	0.127*	0.032	3.925
	6 month	0.839	0.210*	0.031	6.852
	9 month	0.745	0.272*	0.053	5.124
	12 month	0.732	0.297*	0.060	4.952
Japanese yen	3 month	0.709	0.149*	0.032	4.683
	6 month	0.104	0.078	0.076	1.024
	9 month	0.677	0.155*	0.036	4.348
	12 month	0.816	0.200*	0.032	6.311
Australian dollar	3 month	0.634	0.147*	0.037	3.951
	6 month	0.054	0.048	0.067	0.717
	9 month	0.589	0.139*	0.039	3.588
	12 month	0.738	0.175*	0.035	5.030

Note: *indicates significance at 5% level. The results are estimated using the following equation.
$\ln R(j) = \ln \sigma^2 - d \ln 2^{2j}$. Note that the daily volatility is measured by absolute value of the daily change of the LIBOR rates.

to try to model the most short-term interest rates, including LIBOR rates, based on the assumption that the interest rate follows the Brownian motion. Under this assumption, the daily change of the interest rate follows a random walk hypothesis and should not have long-memory behavior. In contrast, our empirical finding clearly shows that the long-memory behaviors of LIBOR rates depend on the currency investigated. Studying the long-memory behaviors of volatilities of LIBOR rates reveals that all volatilities of LIBOR rates possess a long-memory property. This empirical result could be applied to construct the LIBOR market modeling.

4.3. Summary and Concluding Remarks

This chapter investigates the long-memory behavior in rates and volatilities of LIBOR for the four major currencies, namely, the US dollar, the UK pound, the Japanese yen, and the Australian dollar. We propose a new approach — Jensen's (1999a) wavelet OLS estimation technique — to understand this investigation. The method focuses on the estimation of the long-memory parameter for four different maturities and currencies.

As a preliminary test, we plot the autocorrelation coefficients up to 60 lags, implying that the daily percentage changes show little, if any, long-memory behavior. To confirm this result, the WOLS estimates are reported for the long-memory parameter, *d*. We find that the outcome of the test results may vary with different currencies. Overall, the WOLS estimates show that the Japanese yen and Australian dollar LIBOR rates do not show any significant long-memory pattern, while the US dollar and UK pound show significant long-memory pattern regardless of maturities. Finally, for the long-memory behavior for volatilities of LIBOR rates, we find consistently that the volatility has a clear pattern of long memory regardless of maturities and currencies, except for 6-month LIBOR rates for the Japanese yen and Australian dollar.

Chapter 5

Cross-Listing and Transmission of Pricing Information of Dually-Listed Stocks: A New Approach Using Wavelet Analysis

In this chapter, we propose a new approach for investigating how returns and volatilities of dually-listed stocks interact on two non-synchronous Hong Kong and London stocks markets. The proposed method is based on a wavelet multiscale approach that decomposes a given time series on a scale-by-scale basis. Empirical results show that the transmission of information between the HKEx and LSE markets runs in both directions for contemporaneous spillover effects, regardless of the different time scales and trading volumes. Second, for the lagged spillover effects, the impact is stronger moving from the HKEx to the LSE, confirming that trading in the London market plays a limited role in the transfer of pricing information of the Hong Kong market, especially at the shorter time scales. However, at the longer time scales, the evidence shows that the transmission of information runs in both directions, regardless of the different trading volumes.

5.1. Introduction

Due to advanced information technology and increasing internationalization of the capital markets, an increasing number of firms cross-list their shares on both foreign exchanges. The recent popularity of international cross-listings has stimulated much academic research on the topic.[1]

The existing literature on international cross-listings has dealt with a wide range of issues including the motives for listing abroad (Biddle and Saudagaran, 1989; Karolyi, 1998 and Doidge *et al.*, 2004), volume effects in dual-traded stocks (McGuinness, 1999), cross-border listings and

---

[1]Karolyi (1998) provides an extensive survey of international cross-listings.

price discovery (Eun and Sabherwal, 2003), price discovery without trading (Cao *et al.*, 2000), intraday analysis of market integration (Werner and Kleidon, 1996), intraday examination of volume and volatility (Lowengrub and Melvin, 2002), price transmission dynamics between ADRs (Kim *et al.*, 2000), and the bid–ask spread relationship and the return-volatility relationship of dually-listed stocks (e.g., Cheung and Shum, 1995; and Cheung *et al.*, 1995). Despite the large amount of literature on international cross-listings, there is relatively little investigation of the transmission of pricing information between foreign and domestic stock exchanges through internationally-listed securities.

We know of only three published studies and one unpublished work on the transmission of pricing information of dually-listed stocks (Lau and Diltz, 1994; Bae *et al.*, 1999; and Wang *et al.*, 2002; Agarwal *et al.*, 2007). Lau and Diltz (1994) study the transmission of pricing information between the New York Stock Exchange (NYSE) and the Tokyo Stock Exchange (TSE). Using the seemingly unrelated regression (SUR) method, they conclude that market imperfections that may inhibit information transfer between the TSE and the NYSE are not readily apparent and that international listings do not give rise to arbitrage opportunities. Bae *et al.* (1999) investigate the transfer of pricing information using the daily opening and closing prices of eighteen Hong Kong firms that are dually-listed on the Stock Exchange of Hong Kong (HKEx) and the London Stock Exchange (LSE). Also using the SUR approach, they demonstrate that the transmission of information runs in both directions but the stronger impact is from the LSE to the HKEx. Recently, Wang *et al.* (2002) investigate how returns and volatilities of stocks are correlated for dually-listed stocks on two non-synchronous international markets, Hong Kong and London using GJR-GARCH (1, 1) models. They provide evidence of returns and volatility spillovers from Hong Kong to London, and from London to Hong Kong. Agarwal *et al.* (2007) report that if international trading of home market-listed stocks is purely liquidity-driven rather than information-driven, international market prices fully incorporate home market prices but not vice versa. They also use a sample of Hong Kong-listed stocks that are also traded in London.

The objective of this chapter is to investigate the dynamics of the transmission of pricing information in the return and volatility behavior of stocks that are traded in more than one country. Specifically, this chapter employs a new approach (wavelet analysis) for examining the transfer of pricing information of dually-listed stocks using daily opening and closing prices of fourteen Hong Kong firms that are dually-listed on the HKEx and the LSE. The proposed method is based on a wavelet multiscale approach that decomposes a given time series on a scale-by-scale basis. To examine how rapidly the price movements in one market are transmitted to the other market, we use the Granger-causality tests in the wavelet domain for *contemporaneous* and *lagged spillover* effects of daily returns and volatilities.

Our choice of Hong Kong stocks is motivated by several reasons. First, the HKEx represents the largest emerging market in Asia in terms of market capitalization. Second, the LSE trading time does not coincide with the HKEx trading time. A non-overlap of trading hours between two markets in which the same underlying stock is traded provides an ideal opportunity to test information transfer as well as market efficiency (Bae *et al.*, 1999). Third, the London market is the major alternative trading venue of Hong Kong-listed stocks for European and US investors. More interestingly, London trading of Hong Kong-listed stocks has been dominated by European and US institutional investors (Chang, Oppenheimer and Rhee, 1997). Fourth, both the Hong Kong and London markets trade these stocks in HK dollars. Agarwal *et al.* (2007) suggest that this unique feature allows us to get around potential drawbacks which exist in the studies focusing on multi-market trading as identified by Werner and Kleidon (1996). For example, Werner and Kleidon (1996) argue that when stocks are traded in different currency denominations in different markets, there are at least two potential drawbacks: (i) two stocks are not necessarily perfect substitutes for all investors, and (ii) the intraday exchange rate volatility can induce a new risk factor in the pricing of these international stocks.

There are several *innovations* in the chapter. First, to the best of our knowledge, no previous study has investigated the transfer of pricing information of cross-listed stock using a Granger-causality test on the wavelet domain. In this chapter, we employ the Granger-causality test to address the following important question: Does the transmission of pricing information of dually-listed stocks run in one way or in both directions? The current literature does not provide clear findings on the transmission of pricing information: uni- or bi-directional to motivate the study. If we can identify a set of stocks that cause the least amount of confounding complications resulting from private/public information, a straightforward answer may be feasible. The underlying hypothesis is simple: if international trading of stocks traded in multiple markets is purely liquidity-rather than information-driven, international market prices fully incorporate home market prices but not vice versa (Agarwal *et al.*, 2007). A subset of Hong Kong-listed stocks that are also traded in London is an ideal candidate to test this hypothesis. Second, our study extends the Wang *et al.* (2002) using a more recent sample period, October 22, 1996 to December 31, 2002. Wang *et al.* (2002) use 15 dually-listed stocks and an earlier period of October 22, 1996 to July 31, 2000. To examine the transfer of pricing information in a more precise manner, in this chapter, we use the daily stock returns and volatilities of dually-listed stocks instead of using index returns as Wang *et al.* (2002) did in their study. Third, one of the important issues in the transmission of pricing information is how rapidly the price information is transmitted between an emerging market and a developed financial market. To understand the transmission mechanism fully, we use a different (and arguably superior) testing methodology compared with previous studies. The main advantage of wavelet analysis is its ability to decompose the data into several time scales. For example, adopting wavelet analysis allows us to decompose the daily volatilities (unconditional variance) into different time scales so that we can observe which investment horizons are important contributors to the time series volatilities (variance).

Consider the large number of investors who participate in the international stock market and make decisions over *different time scales.* International stock market participants are a diverse group and include intraday traders, hedging strategists, international portfolio managers, commercial banks, large multinational corporations, and national central banks. It is notable that these market participants operate on very different time scales. For example, intraday traders and market makers are looking to both buy and sell a financial assets to realize a quick profit (or minimize a loss) over very short time scales ranging from seconds to hours. Hedging funds strategists often trade over a few days, or on a "close-to-close" basis (Lynch and Zumbach, 2003). Next we have the growing number of international portfolio managers who mainly follow trading strategies, such as index tracking.[2] This typically occurs on a weekly to monthly basis, with little attention paid to intra day prices. Finally, as a market trader, consider the central banks which operate on longer time scales and often consider

---

[2]According to a survey conducted by the Hong Kong Securities and Futures Commission (HKSFC), London market makers indicate that a large proportion of London trades are conducted for portfolio rebalancing under program trading and index fund trading. This indicates that London trades of Hong Kong stocks are more liquidity than information driven (see Chang, Oppenheimer and Rhee, 1997).

long-term economic fundamentals with a very long investment horizon. In short, as a result of the different decision-making time scales among different traders, the true dynamic and causal relationships between international stock markets will vary over different time scales associated with these different investment horizons. Although financial economists have long recognized the idea of several time periods in investment and trading decision making, financial and economic analyses often have been restricted to at most two time scales (the short-run and the long-run), due to the lack of analytical tools to decompose data into more than two time scales.

In the study of the transfer of pricing information of dually-listed stocks, the previous methods are poorly equipped for this task because they use only two time scales (the long run and the short run). For example, studies such as Eun and Sabherwal (2003), and Harris *et al.* (1995) have examined the contribution of cross-listing to price discovery using cointegration (the long-run analysis) and error correction models (the short-run analysis). The recent study by Wang *et al.* (2002) has focused on the transfer of return and volatility behaviors of dually-listed stocks for the case of Hong Kong, using the GJR-GARCH (1,1) models. However, the usual GARCH (1,1) process includes only one time horizon, and this is not enough to replicate the multi-horizon complexity of international stock market trading.

The major innovation of this chapter is the use of wavelet analysis to investigate the causal relationships between dually-listed stocks on an emerging market and a developed market. Unlike previous studies, wavelet analysis allows us to decompose the data into various time scales. This study helps to deepen our understanding of the true dynamics of pricing information transmission over different time scales. The result therefore should be of interest to local intraday traders, hedge managers, international investors, as well as monetary and regulatory authorities, all of whom operate on very different time scales.

Two major empirical findings emerge from wavelet analysis with various time scales: (i) The transmission of information between the HKEx and LSE markets runs in both directions for contemporaneous spillovers effects, regardless of the time scale. (ii) However, for the lagged spillover effects, the impact is stronger moving from the HKEx to the LSE, confirming that trading in the London market plays a limited role in the transfer of pricing information of the Hong Kong market, especially at the shorter time scales. However, at the longer time scales, the evidence shows that the transmission of information runs in both directions.

The remainder of this chapter is organized as follows: Section 5.2 describes the data and provides basic statistics. Section 5.3 presents the empirical results and discusses the findings. The main conclusions and implications are summarized in Section 5.4.

5.2. Data Description and Basic Statistics

In this chapter, we use daily opening and closing prices for the dually-listed stocks on the HKEx and the LSE. Our data are composed of 15 stocks that are actively traded in both the Hong Kong and London markets and are collected from Datastream for the period October 22, 1996 to December 31, 2002.[3] In Fig. 5.1, we present the trading hours of the HKEx and the LSE. The trading hours of the HKEx start at 10:00 a.m. and end at 3:30 p.m. with a two-hour (12:30 to 2:30 p.m.) lunch break, Hong Kong time. Two hours later, the LSE opens at 9:30 a.m. (5:30 p.m. Hong Kong time) and continues until 3:30 p.m. (11:30 p.m. Hong Kong time). Therefore, there is no overlap in trading hours between these two stock exchanges throughout the whole year.

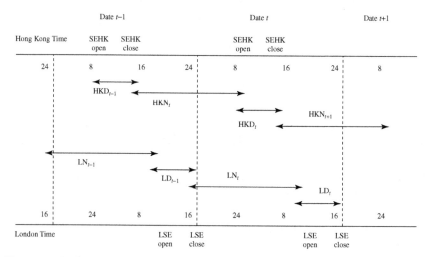

Fig. 5.1. Trading time in the London Stock Exchange and the Stock Exchange of Hong Kong.

---

[3] All companies have the same data period (October 22, 1996 to December 31, 2002) except for three companies: Hang Lung Group (November 11, 1996 to December 31, 2002), New World Development (October 23, 1996 to December 31, 2002) and Wharf Holdings (November 14, 1996 to December 31, 2002)

Let $HKOP_t$ and $HKCP_t$ denote the opening price and closing price on the HKEx on day t, respectively. Then we can define the daytime returns HKD_t and the overnight returns HKN_t on the HKEx as $\log(HKCP_t/HKOP_t)$ and $\log(HKOP_t/HKCP_{t-1})$, respectively. Similarly, let LOP_t and LCP_t be the London opening and closing prices on day t, respectively. We define the daytime returns LD_t and the overnight returns LN_t on the LSE as $\log(LCP_t/LOP_t)$ and $\log(LOP_t/LCP_{t-1})$, respectively.

Fig. 5.1 presents the chronological sequence of the opening and closing times on the HKEx and LSE. The chronological sequence of four returns (HKD_t, HKN_t, LD_t and LN_t) is also presented in Fig. 5.1. As Fig. 5.1 illustrates, there is an overlap in the trading time for LD_{t-1} (LSE daytime returns) and HKN_t (HKEx overnight returns), and HKD_t (HKEx daytime returns) and LN_t (LSE overnight returns). However, HKD_t (HKEx daytime returns) and LD_t (LSE intraday returns) do not overlap.

Table 5.1 provides a summary of basic statistics for the daily daytime return and overnight return series for the 15 dually-listed stocks on the HKEx and the LSE for the sample period from October 22, 1996 to December 31, 2002. From Panel A of Table 5.1, it is not surprising to observe that for all of the 15 dually-listed stocks on the HKEx, their daytime volatilities (measured by standard deviation) are much higher than their overnight volatilities: the daytime return volatility across the 15 dually-listed stocks is 0.026, while the overnight return mean volatility is only 0.007. Furthermore, from Panel B of Table 5.1, we observe that the daytime return mean volatility is 0.009, while the overnight return mean volatility is 0.012 on the LSE. This result is consistent with the findings in Wang *et al.* (2002), Hill *et al.* (1990), and Tse *et al.* (1996). For example, Wang *et al.* (2002) report that daytime trading volatility is greater than overnight non-trading volatility in the HKEx, while the reverse is the case in the LSE. The standard deviations of the daytime return range from 0.015 (Hong Kong Electronics) to 0.031 (Shangri-la Asia) on the HKEx, and from 0.006 (Hong Kong Electronics) to 0.015 (Wharf Holdings) on the LSE. It is worth noting that the London market has a lower variability than the Hong Kong market. Measures of skewness and kurtosis are also reported to indicate whether daytime and overnight returns on the HKEx and the LSE are normally distributed. The sign of skewness and kurtosis varies between markets and firms, confirming that in most cases their empirical distributions have heavy tails relative to the normal distribution. Although we do not report them here, in all cases, the Jarque-Bera statistics reject normality at any conventional level of statistical significance. The Ljung-Box statistics for

Table 5.1.

Panel A. Basic statistics for the dually-listed stocks on the Hong Kong market

		Mean	Std. Dev.	Skewness	Kurtosis	LB(5)[a]	LB(10)	LB^2(5)	LB^2(10)	ρ[b]	$\rho_{Night,Day}$[c]	$\rho(HKN_t, LD_{t-1})$[d]
Cathay Pacific Airways	daytime	0.034	0.028	0.227	2.972	5.945	9.721	116.092*	37.137*	0.032	−0.091	0.210
	overnight	−0.019	0.007	−1.241	12.149	16.801*	24.635*	113.499*	129.457*	0.043		
Cheung Kong Holdings	daytime	−0.061	0.021	0.164	2.493	15.340*	23.662*	85.590*	118.984*	0.065	0.023	0.335
	overnight	0.020	0.007	−0.549	14.709	22.462*	28.415*	213.672*	230.007*	0.006		
Citic Pacific	daytime	−0.114	0.027	0.468	5.209	13.566*	37.531*	129.515*	203.658*	0.026	−0.029	0.254
	overnight	0.021	0.007	−0.356	15.167	14.634*	24.114*	152.862*	169.869*	0.002		
Hong Kong Electronics	daytime	0.020	0.015	0.220	4.372	7.544*	16.613*	217.493*	309.617*	−0.028	−0.037	0.250
	overnight	−0.004	0.004	−1.488	48.471	11.922*	36.780*	46.399*	47.874*	0.007		
Hang Lung Development	daytime	0.039	0.024	0.262	3.092	7.4221*	9.952	204.599*	246.54*	0.068	−0.182	0.263
	overnight	−0.024	0.007	−0.533	5.161	14.5507*	19.046*	75.631*	80.10*	0.064		
Henderson Land	daytime	−0.056	0.024	0.198	1.789	18.634*	23.456*	65.346*	161.324*	0.071	0.117	0.330
	overnight	−0.008	0.006	−0.274	17.011	14.989*	26.529*	231.609*	233.800*	−0.027		
Hang Lung Group	daytime	−0.015	0.023	0.202	4.318	2.438	14.535*	160.664*	212.581*	0.001	−0.001	0.197
	overnight	−0.019	0.006	−1.205	9.414	4.518	15.017*	20.036*	31.934*	0.032		
Hang Seng Bank	daytime	−0.006	0.019	0.364	3.493	8.653*	19.414*	119.226*	426.431*	−0.041	0.088	0.453
	overnight	−0.001	0.006	−1.782	46.448	56.372*	67.347*	269.008*	269.765*	−0.103		

(*Continued*)

Table 5.1.

Panel A. Basic statistics for the dually-listed stocks on the Hong Kong market (Continued)

		Mean	Std. Dev.	Skewness	Kurtosis	LB(5)[a]	LB(10)	LB2(5)	LB2(10)	ρ[b]	$\rho_{Night,Day}$[c]	$\rho(HKN_t, LD_{t-1})$[d]
Hutchison	daytime	-0.042	0.022	0.155	1.922	17.989*	26.196*	175.389*	231.608*	0.003	0.055	0.321
Whampoa	overnight	0.018	0.007	-1.179	21.004	16.713*	26.716*	109.077*	112.021*	-0.016		
Hysan	daytime	-0.106	0.025	0.517	5.037	16.285*	18.180*	140.144*	279.742*	0.066	0.000	0.176
Development	overnight	0.001	0.006	0.151	8.092	5.520	9.314*	66.397*	69.425*	0.009		
New world	daytime	-0.189	0.029	0.337	4.001	14.936*	24.003*	82.931*	128.531*	0.074	0.114	0.272
Development	overnight	0.012	0.008	-0.909	24.511	12.388*	25.568*	40.473*	49.435*	0.013		
Shangri-la	daytime	0.015	0.031	0.218	3.013	6.264*	12.297	211.824*	295.729*	-0.039	-0.061	0.071
Asia	overnight	-0.029	0.007	-0.618	4.482	11.876*	35.318*	16.566*	21.783*	0.069		
Swire	daytime	-0.018	0.025	0.449	3.769	5.989*	10.226	141.900*	5.300	0.023	0.072	0.244
Pacific A	overnight	-0.018	0.006	-1.228	17.757	10.868*	14.085*	19.847*	26.233*	0.039		
Wharf	daytime	-0.089	0.028	0.435	5.105	20.215*	31.616*	136.969*	204.766*	0.102	0.018	0.183
Holdings	overnight	0.014	0.007	0.005	9.871	20.074*	38.900*	304.080*	331.840*	-0.021		
Wheelock &	daytime	-0.020	0.027	0.184	2.280	23.153*	28.685*	148.969*	197.484*	0.116	-0.124	0.241
Co	overnight	-0.027	0.008	0.338	17.727	22.190*	37.587*	206.026*	210.743*	-0.028		

Note: *indicates significance at 5% level. LB(k) and LB2(k) denotes the Ljung-Box test of significance of autocorrelations of k lags for returns and squared returns, respectively. ρ is the first order autocorrelation coefficient. $\rho_{Night,Day}$ is the correlation between daytime return and the preceding overnight return. $\rho(HKN_t, LD_{t-1})$ is the cross correlation between HKEx overnight return and one-period lagged LSE daytime return.

(Continued)

Table 5.1.

Panel B. Basic statistics for the dually-listed Hong Kong stocks on the London stock market

		Mean	Std. Dev.	Skewness	Kurtosis	LB(5)[a]	LB(10)	LB²(5)	LB²(10)	ρ[b]	ρ_{Night,Day}[c]	ρ(LN_t, HKD_t)[d]
Cathay Pacific Airways	daytime	−0.037	0.011	0.409	25.973	12.289*	29.067*	32.359*	37.137*	0.043	−0.091	0.832
	overnight	0.012	0.013	0.030	3.955	12.017*	13.292	125.222*	144.708*	0.024		
Cheung Kong Holdings	daytime	−0.001	0.009	0.927	37.253	12.150*	19.442*	13.928*	17.786*	−0.073	0.023	0.797
	overnight	−0.006	0.011	0.308	6.253	23.823*	27.665*	149.007*	192.925*	0.098		
Citic Pacific	daytime	−0.033	0.009	2.105	45.432	22.529*	26.415*	197.581*	198.994*	−0.095	−0.029	0.850
	overnight	−0.013	0.013	0.335	6.254	17.779*	24.111*	105.932*	155.878*	0.071		
Hong Kong Electronics	daytime	−0.051	0.006	−0.547	10.004	16.724*	19.693*	610.061*	636.593*	−0.013	−0.037	−0.013
	overnight	0.027	0.007	−0.178	7.787	19.049*	32.713*	221.397*	291.269*	−0.040		
Hang Lung Development	daytime	0.084	0.009	0.198	21.152	4.722	16.670*	17.982*	21.542*	0.002	−0.010	0.031
	overnight	−0.044	0.011	−0.001	4.243	10.444*	20.319*	178.868*	223.901*	0.056		
Henderson Land	daytime	−0.008	0.009	−0.877	35.542	18.231*	25.507*	88.335*	97.245*	−0.047	0.117	0.091
	overnight	−0.028	0.012	0.218	2.481	13.465*	15.456*	142.979*	240.886*	0.057		
Hang Lung Group	daytime	0.021	0.009	−1.382	19.838	25.847*	45.773*	142.350*	180.791*	−0.064	−0.001	0.052
	overnight	−0.035	0.011	0.083	5.150	11.297*	19.917*	176.478*	215.695*	0.042		
Hang Seng Bank	daytime	−0.011	0.007	−0.972	20.376	30.587*	38.714*	423.025*	426.431*	−0.107	0.088	0.011
	overnight	0.002	0.010	0.088	4.913	9.970*	18.414*	216.917*	273.256*	−0.067		

(*Continued*)

Table 5.1.

Panel B. Basic statistics for the dually-listed Hong Kong stocks on the London stock market (*Continued*)

		Mean	Std. Dev.	Skewness	Kurtosis	LB(5)[a]	LB(10)	LB^2(5)	LB^2(10)	ρ[b]	ρ$_{Night,Day}$[c]	ρ(LN$_t$, HKD$_t$)[d]
Hutchison	daytime	−0.017	0.009	−1.584	38.648	21.421*	23.959*	32.833*	39.669*	−0.110	0.055	0.026
Whampoa	overnight	.004	0.011	0.034	3.341	11.254*	16.632*	267.291*	378.209*	0.016		
Hysan	daytime	0.002	0.009	0.889	22.547	10.418*	20.075*	189.124*	189.093*	0.003	0.000	0.090
Development	overnight	−0.046	0.012	0.441	7.133	16.976*	19.647*	107.755*	176.694*	0.089		
New world	daytime	−0.025	0.011	−0.935	22.536	9.362*	9.842	149.837*	152.754*	−0.049	0.114	0.122
Development	overnight	−0.058	0.014	0.628	7.236	22.541*	28.745*	85.432*	123.479*	0.102		
Shangri-la	daytime	−0.025	0.011	−1.902	53.661	3.497	4.400	1.898	2.355	−0.037	−0.061	0.786
Asia	overnight	−0.011	0.014	0.283	4.789	24.913*	28.951*	92.600*	112.633*	−0.076		
Swire	daytime	−0.039	0.010	−4.144	84.746	11.138*	21.652*	4.858	5.300	−0.079	0.072	0.867
Pacific A	overnight	−0.008	0.012	0.747	7.735	12.087*	15.442*	46.893*	91.916*	0.041		
Wharf	daytime	0.029	0.015	12.372	240.210	4.616	6.979	0.272	0.312	0.019	0.018	0.787
Holdings	overnight	−0.038	0.014	−0.797	9.821	16.748*	21.989*	25.370*	31.860*	0.102		
Wheelock &	daytime	0.042	0.012	2.833	105.412	26.080*	31.258*	1.194	1.494	−0.099	−0.124	0.750
Co	overnight	−0.054	0.014	−0.258	4.852	7.563*	15.711*	95.790*	110.887*	0.065		

Note: *indicates significance at 5% level. LB(k) and LB2(k) denotes the Ljung-Box test of significance of autocorrelations of k lags for returns and squared returns, respectively. ρ is the first order autocorrelation coefficient. ρ$_{Night,Day}$ is the correlation between daytime return and the preceding overnight return. ρ(LN$_t$, HKD$_t$) is the cross correlation between LSE overnight return and the HKEx daytime return.

k = 5 and k = 10 lags and squared term indicate that significant linear and non-linear dependencies exist.

The sign of the first order autocorrelations of the daytime returns on the HKEx varies between firms, while the daytime returns of most of the stocks on the LSE show significantly negative first order autocorrelation. The results of these different patterns in daytime and overnight volatilities as well as in autocorrelations are consistent with those of Wang *et al.* (1999). Finally, both panels of Table 5.1 also present the cross-autocorrelations between the LSE and the HKEx for the 15 dually-listed stock return series. As expected, we observe significant positive autocorrelations between the HKEx overnight returns and the one-period lagged LSE daytime returns as summarized in the last column of Panel A, and similarly, significant positive correlations between the LSE overnight returns and the HKEx daytime returns as reported in the last column of Panel B. Given the time difference between the HKEx and LSE, this finding leads us to infer that the release of private/public corporate information on the HKEx occurs during the daytime in Hong Kong, which is the overnight time in London Exchange (Wang *et al.*, 2002).

5.3. Empirical Results

This section investigates the transfer of pricing information for dually-listed stocks on two non-synchronous international markets. We use the daily opening and closing prices of 15 Hong Kong firms that are dually-listed in the HKEx and LSE. The use of dually-listed stock returns and volatilities allows us to address the issue of information flows between national stock markets more precisely than using index returns.

Tables 5.2 and 5.3 report the Granger-causality tests of *contemporaneous spillover effects* at different timescales. Table 5.2 presents the results of the Granger causality test in the wavelet domain for the contemporaneous spillover effects of daily returns (Panel A) and daily volatilities (Panel B) between HKN_t (Hong Kong overnight) and LD_{t-1} (London daytime). We report and interpret the results of Granger causality test between Hong Kong market and London market at different timescales, which covers the time scale from d1 to d7. For example, note that the first shortest wavelet scale d1 captures oscillation with a period length 2 to 4 days while the longest wavelet scale d7 captures oscillation with a period length of 128 to 256 days. Panel A of Table 5.2 shows that the impact of the LSE daytime returns on the HKEx overnight returns is strong, evidenced by

Table 5.2.   Granger causality test in wavelet domain.

Panel A. Contemporaneous spillover effects of daily returns between HKN_t and LD_{t-1}

Size quintile	Causal direction	d1	d2	d3	d4	d5	d6	d7
All	London → Hong Kong	13	9	13	8	12	12	11
	Hong Kong → London	12	10	12	8	10	10	10
Large Companies	London → Hong Kong	5	2	5	3	4	4	4
	Hong Kong → London	5	4	5	3	3	4	5
Medium Companies	London → Hong Kong	3	3	3	2	4	4	4
	Hong Kong → London	2	2	3	2	3	3	3
Small Companies	London → Hong Kong	5	4	5	3	4	4	3
	Hong Kong → London	5	4	4	3	4	3	2

Note: In this table, the original data has been transformed by the wavelet filter [LA(8)] up to time scale 7 and the first detail (wavelet coefficient) d1 captures oscillation with a period length 2 to 4 days. The last detail d7 captures oscillation with a period length of 128 to 256 days. London→ Hong Kong indicates the Granger causality test in wavelet domain for the contemporaneous spillover effects of daily *returns* from LD_{t-1} to HKN_t, and Hong Kong → London indicates the Granger causality test in wavelet domain for the contemporaneous spillover effects of daily *returns* from HKN_t to LD_{t-1}.

Panel B. Contemporaneous spillover effects of daily volatilities between HKN_t and LD_{t-1}

Size quintile	Causal direction	d1	d2	d3	d4	d5	d6	d7
All	London → Hong Kong	8	7	10	8	10	9	10
	Hong Kong → London	12	7	9	9	11	15	9
Large Companies	London → Hong Kong	3	3	4	4	5	4	5
	Hong Kong → London	5	4	5	4	5	5	5
Medium Companies	London → Hong Kong	4	1	3	2	3	2	4
	Hong Kong → London	4	1	2	2	4	5	2
Small Companies	London → Hong Kong	1	3	3	2	2	3	1
	Hong Kong → London	3	2	2	3	2	5	2

Note: In this table, the original data has been transformed by the wavelet filter [LA(8)] up to time scale 7 and the first detail (wavelet coefficient) d1 captures oscillation with a period length 2 to 4 days. The last detail d7 captures oscillation with a period length of 128 to 256 days. London → Hong Kong indicates the Granger causality test in wavelet domain for the contemporaneous spillover effects of daily *volatilities* from LD_{t-1} to HKN_t, and Hong Kong → London indicates the Granger causality test in wavelet domain for contemporaneous spillover effects of daily *volatilities* from HKN_t to LD_{t-1}.

Table 5.3.   Granger causality test in wavelet domain.

Panel A. Contemporaneous effects of daily returns between HKD_t and LN_t

Size quintile	Causal direction	d1	d2	d3	d4	d5	d6	d7
All	London → Hong Kong	8	8	10	12	10	11	12
	Hong Kong → London	9	10	10	11	9	11	12
Large Companies	London → Hong Kong	4	4	5	5	5	5	3
	Hong Kong → London	5	5	5	5	5	5	2
Medium Companies	London → Hong Kong	2	1	2	4	2	3	4
	Hong Kong → London	2	1	2	3	1	3	5
Small Companies	London → Hong Kong	2	3	3	3	3	3	5
	Hong Kong → London	2	4	3	3	3	3	5

Note: In this table, the original data has been transformed by the wavelet filter [LA(8)] up to time scale 7 and the first detail (wavelet coefficient) d1 captures oscillation with a period length 2 to 4 days. The last detail d7 captures oscillation with a period length of 128 to 256 days. London → Hong Kong indicates the Granger causality test in wavelet domain for the contemporaneous spillover effects of daily *returns* from LN_t to HKD_t, and Hong Kong → London indicates the Granger causality test in wavelet domain for the contemporaneous spillover effects of daily *returns* from HKD_t to LN_t.

Panel B. Contemporaneous effects of daily volatilities between HKD_t and LN_t

Size quintile	Causal direction	d1	d2	d3	d4	d5	d6	d7
All	London → Hong Kong	10	11	11	11	14	9	14
	Hong Kong → London	10	11	9	10	14	10	13
Large Companies	London → Hong Kong	5	5	5	5	5	3	4
	Hong Kong → London	4	5	4	4	5	3	4
Medium Companies	London → Hong Kong	2	3	4	3	4	2	5
	Hong Kong → London	3	2	3	2	4	3	4
Small Companies	London → Hong Kong	3	3	2	3	5	4	5
	Hong Kong → London	3	4	2	4	5	4	5

Note: In this table, the original data has been transformed by the wavelet filter [LA(8)] up to time scale 7 and the first detail (wavelet coefficient) d1 captures oscillation with a period length 2 to 4 days. The last detail d7 captures oscillation with a period length of 128 to 256 days. London → Hong Kong indicates the Granger causality test in wavelet domain for the contemporaneous spillover effects of daily *volatilities* from LN_t to HKD_t, and Hong Kong → London indicates the Granger causality test in wavelet domain for the contemporaneous spillover effects of daily *volatilities* from HKD_t to LN_t.

the fact that 13 out of 15 firms at the finest wavelet scale d1, 9 out of 15 firms at the second wavelet scale d2 and 11 out of 15 firms at the longest wavelet scale d7 are statistically significant. Similarly, the impact of the HKEx overnight returns on the LSE daytime returns is equally strong. For example, from Panel A of Table 5.2, we observe that 12 out of 15 firms (d1), 10 out of 15 firms (d2), and 10 out of 15 firms (d7) are statistically significant. Hence, the LSE returns also respond rapidly to changes in returns on the HKEx. From the columns of Panel B of Table 5.2 the impact of the LSE daytime volatilities on the HKEx overnight volatilities are also statistically significant (i.e., 8 for d1, 7 for d2 and 10 for d7 time scale). In reverse, the impact of the HKEx overnight volatilities on the LSE daytime volatilities is strong and statistically significant (i.e., 12 for d1, 7 for d2 and 9 for d7 time scale). Given the sample size of 15 firms, it would be interesting to examine how the volume (size) affects the overall findings-for example, how do the results differ among those dually-listed stocks with large, medium and small trading volume in the exchanges. To address this issue, in appendix we present three different groups of firms which are classified by the market values of each company using a benchmark year of 2002. From the Panel A and B of Table 5.2, the results of the Granger causality test at different timescales for three different firm groups (i.e., large, medium and small companies) indicate that there are no significant volume effects across different firm size. From both of Panel A and B, it is shown that the trading volume (firm size) effects are not much different among three different firm groups, regardless of the different time scale. In short, from Table 5.2 we observe that contemporaneous spillover effects of daily returns and volatilities between HKN_t (Hong Kong overnight) and LD_{t-1} (London daytime) are bi-directional and significant, regardless of the different time scales and trading volumes.

Table 5.3 presents the results of the Granger causality test in the wavelet domain the contemporaneous spillover effects of daily returns (Panel A) and daily volatilities (Panel B) between HKD_t (Hong Kong daytime) and LN_t (London overnight). The interpretation of the results in Table 5.3 is very similar to that of Table 5.2. Therefore, the key common finding from Tables 5.2 and 5.3 is that, overall, the effect from the HKEx to the LSE is equally significant to the effect from the LSE to the HKEx, regardless of the different time scales and trading volumes.

In short, the main empirical finding implies that the bilateral transmission relationship reported here using dually-listed stocks provides clear evidence that stock returns and volatilities on the emerging market (Hong Kong) also rapidly affect the corresponding returns and volatilities in the developed market (London). This empirical finding is consistent with that of Bae *et al.* (1999).

To investigate the *lagged spillover effects* between two markets, Tables 5.4 and 5.5 present the results of the Granger causality test in the

Table 5.4. Granger causality test in wavelet domain: Lagged spillover effects of daily returns.

Size quintile	Causal direction	d1	d2	d3	d4	d5	d6	d7
All	London → Hong Kong	4	7	8	9	9	13	12
	Hong Kong → London	14	13	12	14	14	15	14
Large Companies	London → Hong Kong	1	2	4	2	3	5	4
	Hong Kong → London	5	5	4	5	4	5	5
Medium Companies	London → Hong Kong	2	1	2	4	4	4	5
	Hong Kong → London	5	4	4	4	5	5	5
Small Companies	London → Hong Kong	1	4	2	3	2	4	3
	Hong Kong → London	4	4	4	5	5	5	4

Note: In this table, the original data has been transformed by the wavelet filter [LA(8)] up to time scale 7 and the first detail (wavelet coefficient) d1 captures oscillation with a period length 2 to 4 days. The last detail d7 captures oscillation with a period length of 128 to 256 days. London → Hong Kong indicates the Granger causality test in wavelet domain for the lagged spillover effects of daily *returns* from LD_{t-1} to HKD_t, and Hong Kong → London indicates the Granger causality test in wavelet domain for the lagged spillover effects of daily *returns* from HKD_t to LD_t.

wavelet domain from LD_{t-1} to HKD_t and from HKD_t to LD_t for daily returns (Table 5.4) and daily volatilities (Table 5.5), respectively. The first important finding in Tables 5.4 and 5.5 is the unidirectional influence in the time scale of d1, d2, and d3 series of the dually-listed stocks between the HKEx and the LSE, as indicated by the evidence that the causal direction from Hong Kong trading to the London market is significant for 14 out of 15 firms for d1, 13 for d2, and 12 for d3 time scale for daily returns (Table 5.4) and 13 for d1, 14 for d2, and 13 for d3 timescale for daily volatilities (Table 5.5). An important finding is that the an unidirectional influence of causality effects from the Hong Kong market to the London market is consistently strong for both of daily returns and volatilities, regardless of the different time scales and trading volumes. To the contrast, the dissemination of pricing information from London trading to the Hong Kong market is relatively weak, as indicated by the evidence that the causality effects of daily returns direction from the LSE to the HKEx is less significant. For example, there is only 4 out of 15 firms for d1, 7 for d2, and 8 for d3 time scale for daily returns (Table 5.4) and only 5 for d1, 6 for d2 and 9 for d3 time scale for daily volatilities (Table 5.5), confirming that trading in the London market plays a limited role in price discovery in the Hong Kong market.

Overall, however, the causality effects from LSE market to the HKEx market are displayed strongly beyond the third layer, which represents a

Table 5.5. Granger causality test in wavelet domain: Lagged spillover effects of daily volatilities.

Size quintile	Causal direction	d1	d2	d3	d4	d5	d6	d7
All	London → Hong Kong	5	6	9	8	7	9	12
	Hong Kong → London	13	14	13	14	14	15	14
Large Companies	London → Hong Kong	3	2	3	4	2	4	3
	Hong Kong → London	5	5	4	5	4	5	5
Medium Companies	London → Hong Kong	1	2	3	3	3	0	5
	Hong Kong → London	4	4	4	4	5	5	4
Small Companies	London → Hong Kong	1	2	3	1	2	5	4
	Hong Kong → London	4	5	5	5	5	5	5

Note: In this table, the original data has been transformed by the wavelet filter [LA(8)] up to time scale 7 and the first detail (wavelet coefficient) d1 captures oscillation with a period length 2 to 4 days. The last detail d7 captures oscillation with a period length of 128 to 256 days. London → Hong Kong indicates the Granger causality test in wavelet domain for the lagged spillover effects of daily *volatilities* from LD_{t-1} to HKD_t, and Hong Kong → London indicates the Granger causality test in wavelet domain for the lagged spillover effects of daily *volatilities* from HKD_t to LD_t.

data length of around 8 days. It should be noted that the two stock markets become more strongly related as the time-scale increases. In other word, the wavelet multiscaling Granger-causality tests indicate that the transfer of pricing information between two markets in terms of daily returns and volatilities is more certain in the long term, regardless of the different trading volumes.

5.4. Concluding Remarks

In this chapter, we investigate the issue of pricing information transfer on two non-synchronous international markets. We propose a new approach based on the wavelet multiscaling method to test for dependencies and the direction of spillover effects. The approach focuses on the important question: Does the transmission of pricing information of the dually-listed stocks run in one way or in both directions?

To examine how rapidly the price movements in one market are transmitted to the other market, we employ the Granger-causality test for various time scales on dually-listed stocks using wavelet analysis. Using a sample of 15 Hong Kong stocks listed on the HKEx that are also listed in the UK on the LSE, we find that there is some evidence that the transfer of pricing information is bi-directional, since the Granger causality tests in the wavelet domain for *contemporaneous spillover* effects of daily returns and volatilities run in both directions. In other words, our empirical evidence indicates that Hong Kong stocks are priced to reflect information from the London market as well as the Hong Kong market. This evidence of bi-directional causality effects is consistent for the different time scales and trading volumes.

However, Granger-causality tests for *lagged spillover* effects of daily returns and volatilities run in one direction and show that the return and volatility spillovers from the HKEx to the LSE are much stronger than those from the LSE to the HKEx, especially at the fine wavelet scales d1, d2 and d3. The dissemination of return and volatility spillovers from the HKEx to the LSE is consistently strong, regardless of the different time scales and trading volumes. In other words, for the lagged spillover effects, the impact is stronger moving from the HKEx to the LSE, confirming that trading in the London market plays a limited role in the transfer of pricing information of the Hong Kong market, especially at the shorter time scales. However, at the longer time scales, the evidence shows that the transmission of information runs in both directions, regardless of the different trading volumes.

Appendix. The market values of sample companies as of 2002

The data are obtained from Datastream. The market value for 2002 is calculated as sum of the daily market value of each firm during 2002.

Group	Name	Market Value	Total
Large Company	Hutchison Whampoa	66,282,731.79	
	Hang Seng Bank	42,701,948.45	
	Cheung Kong Holdings	38,367,242.79	
	Hong Kong Electronics	16,655,388.87	
	Handerson Land	13,517,577.68	177,524,889.58
Medium Company	Wharf Holdings	10,970,558.65	
	Cathay Pacific Airways	10,107,132.47	
	Citic Pacific	9,383,863.71	
	Swire Pacific A	9,330,949.01	
	Hang Lung Props	5,988,296.84	45,780,800.68
Small company	Shangri-la Asia	3,284,122.75	
	New World Development	3,200,238.51	
	Wheelock & Co.	3,165,675.71	
	Hang Lung Group	2,405,713.93	
	Hysan Development	1,959,348.82	14,015,099.72

Chapter 6

On the Relationship Between Stock Returns and Risk Factors: New Evidence From Wavelet Analysis

This chapter adopts a new approach to examine the Fama–French three-factor model, which is commonly used to explain the cross-sectional variation in average stock returns over various times. The new approach is based on the wavelet multiscaling method that decomposes a given time series on a scale-by-scale basis. The empirical results reveal that if the mispricing is correct, it is expected that the loadings on the HML will be significant in the short scales, while insignificant in the long scales. It is found that all loading on the HML is significant regardless of time scales and portfolios. It is also found that the excess market return and HML combine to capture the cross-sectional variation in the six size-BM portfolio returns, and that SMB is effective in explaining the cross-section, except for large stocks.

6.1. Introduction

Although the traditional Capital Asset Pricing Model (CAPM) plays an important role in the way academics and practitioners think about risk and the relationship between risk and return, the CAPM is not successful in the recent period. The failure of the CAPM to explain the cross-sectional variation of expected stock returns has lead many researchers to look for other factors to help predict average stock returns. The list of empirically determined average stock return factors includes firm characteristics such as size (ME, stock price times number of shares), earnings/price (E/P), leverage, cash flow/price (C/P), book-to-market equity (B/M, the ratio of the book value of a common stock to its market value), past sales growth, short-term past return, and long-term past return (Fama and French, 1992). Since these predictable factors in average stock returns are not explained by the CAPM, they are typically regarded as anomalies (Fama and French, 1996).

Most previous studies examine the relationship between stock returns and the risk factors at the short horizon. Hence, previous research has often presented a limited understanding of the true dynamic relationship between stock returns and risk factors, due to the limited time scale.

The importance of multiscaling approach can be found in several previous studies. For example, Levhari and Levy (1977) show that the beta estimate are biased, if the analyst uses a time horizon shorter than the true time horizon, defind as the relevant time horizon implicit in the decision making process of investors (Gencay *et al.*, 2005). Handa *et al.* (1989) report that if we consider different return interval, different betas can be estimated for the same stock. Along with these aspects, Kothari and Shanken (1998) examine the Fama–French model and conclude that Fama and French's results hinge on using monthly rather than yearly returns. Kothari and Shanken (1998) argue that the use of annual returns to estimate betas helps to circumvent measurement problems caused by non-synchronous trading, seasonality in returns, and trading frictions. The arguments of these papers may arise because long horizon traders will essentially focus on price fundamentals that drive overall trends, whereas short-term traders will primarily react to incoming information within a short-term horizon[1] (Connor and Rossiter, 2005). Hence, market dynamics in the aggregate will be the result of the interaction of agents with heterogeneous time horizons. From their point of view, important and interesting questions arise whether expected stock returns correspond differently to risk factors over different time horizons.[2]

---

[1]Connor and Rossiter (2005) point out this argument for the specific case of commodity markets. However, we consider this argument could be applicable to most financial markets, especially in stock market.

[2]The logic is as follows. Financial markets are made of investors and traders with different investment time horizons. In the heart of the trading mechanisms are the market makers. A next level up is the intraday investors who carry out trades only within a given trading day. Then there are day investors who may carry positions overnight, short-term traders and long-term traders. Overall, it is the sum of the activities of all investors for all different investment time horizons that generates the market prices. Therefore, market activity is heterogenous with each investment horizon dynamically providing feedback across the different time scales (Dacorogna *et al.*, 2001). The implication of a heterogenous market is that the true dynamic relationship between the various aspects of market activity will only be revealed when the market prices are decomposed by the *different time scales* or *different investment horizons*.

The main purpose of this chapter is to examine whether or not the Fama–French three factor model holds over the different time scales. To do so, we introduce a new approach: wavelet analysis. The new approach is based on the wavelet multiscaling method that decomposes a given time series on a scale-by-scale basis. The main advantage of wavelet analysis[3] is the ability to decompose the data into several time scales. Consider the large number of investors who trade in the security market and make decisions over different time scales. One can visualize traders operating minute-by-minute, hour-by-hour, day-by-day, month-by-month, and year-by-year. In fact, due to the different decision-making time scales among traders, the true dynamic structure of the relationship between stock returns and risk factors will *vary* over different time scales associated with those different horizons.

When we examine the performance of the three risk factors over various time scales, another testable hypothesis is the mispricing. Lakonishok *et al.* (1994) argue that the book-to-market effect reflects investor's over-reaction rather than compensation for risk bearing. According to them, investors systematically overreact to recent corporate news, and naively extrapolate past earnings growth into the future when evaluating a firm's prospects. Therefore, growth stocks (low book-to-market stocks) are overpriced, while value stocks (high book-to-market stocks) are underpriced. In addition, Daniel *et al.* (1998) assume that investors are overconfident about their private information and over react to it. The increase in overconfidence furthers the initial overreaction and generates return momentum.[4] The overreaction in prices will eventually be corrected in the long-run as investors observe future news and realize their errors (Cooper *et al.*, 2004). Thus, the mispricing could be considered as a short term phenomenon. If this argument is correct, it is expected that the loadings on the HML is significant in the short scales, while insignificant in the long scales.

---

[3]Wavelet analysis is relatively new in economics and finance, although the literature on wavelets is growing rapidly. To the best of our knowledge, applications in these fields include examination of foreign exchange data using waveform dictionaries (Ramsey and Zhang, 1997); examining the relationship between financial variables and industrial production (Kim and In, 2003); decomposition of economic relationships of expenditure and income (Ramsey and Lampart, 1998a, 1998b); the multiscale relationship between stock returns and inflation (Kim and In, 2005b); scaling properties of foreign exchange volatility (Gençay *et al.*, 2001); systematic risk in a capital asset pricing model (Gençay *et al.*, 2003, 2005); and the multiscale hedge ratio (In and Kim, 2006).

[4]Hong and Stein (1999) also develop a behavioral theory, based on initial underreaction to information and subsequent overreaction, which eventually leads to stock price reversal in the long run.

To the best of our knowledge, this chapter is the *first* to investigate the three-factor model and a momentum factor using wavelet analysis. This study helps to deepen our understanding of the true relationship between stock returns and the four risk factors over different time scales. The results therefore should be of interest to both international and local investors. To examine our purpose, we add one factor by one factor into the traditional CAPM. More specifically, first, we examine the CAPM and investigate the difference between the different time scales. To examine the role of SMB, we add SMB into the traditional CAPM. After analyzing the role of SMB, the risk factor, HML, is added to construct the Fama and French three factor model.

This chapter is organized as follows. Section 6.2 presents the data and basic statistics. Section 6.3 discusses the empirical results. We conclude in Section 6.4 with a summary of our results.

6.2. Data and Basic Statistics

We use monthly six size-BM portfolio returns, the excess market return, SMB and HML for the US from January 1927 to December 2003, obtained from the Kenneth French homepage. More specifically, at the end of June each year, all stocks are grouped on the median market capitalization of all NYSE, AMEX and NASDAQ stocks into two groups: small (S) and big (B). Stocks are also independently sorted into three BM groups: low (L), medium (M) and high (H), where L, M and H represent the bottom 30%, middle 40% and top 30% of stocks, respectively. BM is the ratio of book value of equity to market value of equity of a firm for the fiscal year ending in year $t - 1$. Six size-BM portfolios (S/L, S/M, S/H, B/L, B/M, B/H) are defined as the intersections of the two size and three BM groups. The monthly value-weighted average returns of each portfolio are then computed. SMB is the difference between the average returns of the three small-stock portfolios (S/L, S/M, S/H) and three big-stock portfolios (B/L, B/M, B/H). HML is the difference in returns between the two high-BM portfolios (S/H and B/H) and two low-BM portfolios (S/L and H/L). The excess market return is constructed by the difference between the value-weight return on all NYSE, AMEX, and NASDAQ stocks (from CRSP) and the one-month Treasury bill rate. Table 6.1 presents several summary statistics for the monthly data of six size-BM portfolios, the excess market returns, SMB and HML.

Table 6.1.   Basic statistics.

Panel A. Descriptive statistics

	S/L	S/M	S/H	B/L	B/M	B/H	MKT	SMB	HML
Mean	0.467	0.889	1.070	0.419	0.522	0.670	0.449	0.271	0.427
Std.dev.	7.140	5.419	5.440	4.838	4.322	4.491	4.518	3.290	2.992
Skewness	−0.330	−0.537	−0.335	−0.291	−0.265	−0.132	−0.491	0.507	0.029
Kurtosis	1.746	3.337	4.272	1.766	2.105	1.827	1.942	5.382	2.344
LB(5)	15.182*	20.307*	23.317*	3.923	12.926*	7.014*	5.656	7.379*	12.064*
LB(10)	19.540*	29.269*	32.721*	5.575*	18.386*	12.718*	8.858	11.642	14.794*
LB²(5)	13.825*	2.187	1.259	17.944*	2.138	5.615	9.135*	131.151*	230.344*
LB²(10)	17.498*	3.988	1.805	24.715*	12.266	8.659	15.791*	131.440*	330.359*
ρ	0.158	0.188	0.198	0.056	0.012	0.031	0.058	0.072	0.133

Panel B. Correlation matrix with the risk factors

	S/L	S/M	S/H	B/L	B/M	B/H
MKT	0.8681	0.8683	0.8295	0.9687	0.9211	0.8596
SMB	0.6993	0.6664	0.6517	0.2043	0.1269	0.1657
HML	−0.5283	−0.2659	−0.1156	−0.5386	−0.1576	0.0525

Note: The table shows the basic statistics for S/L, S/M, S/H, B/L, B/M, B/H, excess market return (MKT), SMB, HML and UMD. The sample period ranges from January 1927 to December 2003, obtained from the Kenneth French homepage. To construct six size-BM portfolios. all stocks are grouped on the median market capitalization of all NYSE, AMEX and NASDAQ stocks into two groups: small (S) and big (B). Stocks are also independently sorted into three B/M groups: low (L), medium (M) and high (H), respectively. BM is the ratio of book value of equity to market value of equity of a firm for the fiscal year ending in year $t − 1$. Six size-BM portfolios (S/L, S/M, S/H, B/L, B/M, B/H) are defined as the intersections of the two size and three BM groups. SMB is the difference between the average returns of the three small-stock portfolios (S/L, S/M. S/H) and three big-stock portfolios (B/L, B/M, B/H). HML is the difference in returns between the two high-BM portfolios (S/H and B/H) and two low-BM portfolios (S/L and H/L). * indicates significance at 5% level. LB(k) and LB²(k) denotes the Ljung-Box test of significance of autocorrelations of k lags for returns and squared returns, respectively. ρ is the first order autocorrelation coefficient. Skewness and kurtosis are defined as $E[(R_t − \mu)]^3$ and $E[(R_t − \mu)]^4$, where μ is the sample mean.

As shown in Panel A in Table 6.1, all sample means are positive and range from 0.271 (SMB) to 1.070 (S/H). The average return of excess market return is 0.449, of SMB is 0.271, of HML is 0.427. Since average return for each factor is positive investors are compensated from a positive premium for bearing factor risk. From the standard deviation of six portfolios, it is observed that the small stocks are more volatile than the large stocks, while the returns on the small stocks are higher than those of the large stocks, implying that investors want more compensation for the higher risk stocks. Among six size-BM portfolios, the excess market returns, SMB and HML, first-order autocorrelation of monthly data ranges from 0.012 (B/M) to 0.198 (S/H). Furthermore, it is observed that the first-order autocorrelation coefficients of the small stocks are slightly bigger than those of the large stocks, implying that the small stocks are slightly more persistent than the large stocks. The measures of skewness and kurtosis are also reported to indicate whether our data (the six size-BM portfolio return, MKT, SMB and HML) are normally distributed. The signs of skewness and kurtosis vary depending on the portfolio returns, confirming that in most cases their empirical distributions have heavy tails relative to the normal distribution. Although we do not report them here, in all cases, the Jarque-Bera statistics reject normality at any conventional level of statistical significance. The Ljung-Box statistics for $k = 5$ and $k = 10$ lags and squared term indicate that significant linear and non-linear dependencies exist.

Panel B of Table 6.1 presents the sample correlation between the six size-BM portfolio returns and the four risk factors. Overall, the six portfolio returns show higher correlation with the MKT than with the other risk factors. The SMB, the difference between the return on a portfolio of small stocks and the return on a portfolio of large stocks, shows higher correlation with small stocks than large stocks, as expected from the calculation of SMB. This tendency is also observed in the correlation with HML, the difference between the return on a portfolio of high book-to-market stocks and the return on a portfolio of low book-to-market stocks. Two portfolios (S/H and B/H) show higher correlation with HML.

6.3. Empirical Results

It is important to examine the relationship between the risk factors and the portfolio returns over different time scales to reflect the various investment horizons of investors. In this section, applying the time series regression analysis, we present the empirical results for the three-factor model utilizing the decomposed data, and then extend our analysis to include the four factors.

6.3.1. *Results from the traditional CAPM*

This sub-section examines the multiscale relationship between stock returns and excess market returns in the wavelet domain across various time scales. In our investigation, we run the following time series regression using the wavelet coefficients for scale $\lambda_j \equiv 2^{j-1}$ where $j = 1, 2, \ldots, 5$:

$$R_{it}(\lambda_j) - R_{ft}(\lambda_j) = \alpha(\lambda_j) + \beta_{MKT}(\lambda_j)MKT_t(\lambda_j) + \varepsilon_{i,t}(\lambda_j) \qquad (6.1)$$

where $R_{it}(\lambda_j)$ is the return on portfolio i in calendar month t at scale λ_j; $R_{ft}(\lambda_j)$ is the risk-free return (one-month Treasury bill) in calendar month t at scale λ_j; $MKT_t(\lambda_j)$ is the excess market returns, measured by $R_{mt}(\lambda_j)$ minus $R_{ft}(\lambda_j)$, where $R_{mt}(\lambda_j)$ is the CRSP value-weighted market index return in calendar month t at scale λ_j; intercept $\alpha(\lambda_j)$ in equation (6.1) is the abnormal return of portfolio i at scale λ_j, and $\beta_{MKT}(\lambda_j)$ is the assigned loadings on the market at scale λ_j.

We report the estimated coefficients of $\alpha(\lambda_j)$, $\beta(\lambda_j)$, and R^2 in Table 6.2. To generate asymptotically valid standard errors, we report the heteroskedasticity adjusted error using Newey and West's (1987) method to insure the variance-covariance matrix is positive definite. The first row of each dependent variable (six size-BM portfolios) in Table 6.2 presents the results of the original data set for comparison. Note that these results are not from the wavelet multiscaling transform. For the six size-BM portfolios we examine, the intercepts from the CAPM regressions, which include only the excess market return, are significantly different from 0 at S/M, S/H, and B/H portfolios. Thus, a market factor does not seem to reasonably explain the cross-sectional variation in average stock returns.

We now turn to the results of the regression analysis focusing on the different time scales. Considering the sample size and the length of the wavelet filter, we settle on the MODWT based on the Daubechies least asymmetric wavelet filter of length 8 [LA(8)], while our decompositions go to scale 5 (equivalent to the 32–64 month period). We examine the R^2s of regressions, which appear to show that the CAPM has significant explanatory power in cross-sectional variation in average stock returns in

Table 6.2.   Estimated coefficients of the CAPM model on the wavelet domain.

		Raw data	Scale 1	Scale 2	Scale 3	Scale 4	Scale 5
S/L	α	−0.1489	0.0000	0.0013	0.0017	−0.0140	0.0242
		(0.1648)	(0.0412)	(0.0500)	(0.0784)	(0.0709)	(0.0610)
	β_{MKT}	1.3719*	1.2349*	1.4585*	1.6458*	1.5355*	1.2813*
		(0.0402)	(0.0481)	(0.0514)	(0.0553)	(0.0457)	(0.0686)
	R^2	0.7532	0.7098	0.7836	0.8505	0.8743	0.8733
S/M	α	0.4216*	−0.0007	0.0021	−0.0127	−0.0045	0.0097
		(0.1398)	(0.0303)	(0.0344)	(0.0583)	(0.0498)	(0.0512)
	β_{MKT}	1.0415*	0.9109*	1.1335*	1.2719*	1.2294*	1.0325*
		(0.0432)	(0.0466)	(0.0417)	(0.0359)	(0.0374)	(0.0423)
	R^2	0.7535	0.7052	0.8142	0.8681	0.9004	0.8711
S/H	α	0.6213*	−0.0022	0.0033	−0.0157	−0.0062	0.0130
		(0.1612)	(0.0331)	(0.0418)	(0.0771)	(0.0582)	(0.0697)
	β_{MKT}	0.9988*	0.8724*	1.1073*	1.2344*	1.0857*	0.9115*
		(0.0492)	(0.0520)	(0.0474)	(0.0519)	(0.0426)	(0.0542)
	R^2	0.6875	0.6528	0.7567	0.7853	0.8353	0.7414
B/L	α	−0.0468	−0.0004	−0.0002	0.0037	−0.0042	−0.0035
		(0.0606)	(0.0135)	(0.0174)	(0.0322)	(0.0386)	(0.0307)
	β_{MKT}	1.0373*	1.0433*	1.0249*	1.0523*	1.0258*	1.0570*
		(0.0158)	(0.0196)	(0.0189)	(0.0199)	(0.0260)	(0.0225)
	R^2	0.9382	0.9413	0.9403	0.9385	0.9125	0.9520
B/M	α	0.1266	0.0002	−0.0008	−0.0067	0.0082	0.0079
		(0.0904)	(0.0193)	(0.0229)	(0.0416)	(0.0305)	(0.0258)
	β_{MKT}	0.8810*	0.9073*	0.8896*	0.7969*	0.9157*	0.9487*
		(0.0286)	(0.0324)	(0.0276)	(0.0256)	(0.0225)	(0.0233)
	R^2	0.8482	0.8593	0.8649	0.8465	0.9264	0.9572
B/H	α	0.2862*	0.0004	−0.0040	−0.0111	0.0004	0.0002
		(0.1141)	(0.0288)	(0.0316)	(0.0567)	(0.0468)	(0.0495)
	β_{MKT}	0.8544*	0.8618*	0.8558*	0.8501*	0.8233*	0.8285*
		(0.0390)	(0.0446)	(0.0350)	(0.0409)	(0.0424)	(0.0488)
	R^2	0.7384	0.7311	0.7633	0.7642	0.8177	0.8250

Note: The table reports the estimated coefficients of the CAPM model on the wavelet domain for six size-BM portfolios. Data used are monthly US six size-BM portfolios and excess market return for the period January 1927 to December 2003. Data were obtained from the Kenneth French homepage. The heteroskedasticity adjusted error using Newey and West's (1987) method are in parentheses. * indicates the significance at 5% level. The wavelet coefficients are calculated using the Daubechies least asymmetric wavelet filter of length 8 [LA(8)] up to time scale 5. The scale 1 captures oscillation with a period length 2 to 4 months. The last scale 5 captures oscillation with a period length of 32 to 64 months.

the US. Overall, all values of R^2 increase as the time scale increases. In addition, all portfolio returns have the highest R^2 in the high time scale (scales 4 and 5). This result implies that the explanatory power of the CAPM is increasing with time scale. This is a similar result with Gencay *et al.* (2003),[5] who conclude that the predictions of the CAPM are more relevant in the medium long run as compared to short horizons.

From the results of the regression analysis on wavelet time domain, presented in Table 6.2, two things are worth noting. Firstly, the intercepts are not significantly different from 0 at six size-BM portfolios regardless of time scales. Considering the scaling coefficients capture the underlying smooth behavior of time series, while the wavelet coefficients represent the deviations from the smooth behavior, the abnormal return may occur at smoothing trends of portfolio returns.

Secondly, the excess market return seems to play a role in explaining the cross-section of average stock returns regardless of time scale, evidenced by the fact that all market coefficients, $\beta_{MKT}(\lambda_j)$, are statistically significant in all wavelet time scales. However, we find that there exists a size effect. Generally, $\beta_{MKT}(\lambda_j)$s of small stocks are higher than those of big stocks. To examine this relationship between the betas and average returns of six size-BM portfolios, Fig. 6.1 is plotted. Figure 6.1 illustrates average monthly returns of six size-BM portfolios (vertical axis) and corresponding betas at different wavelet scales.

From Fig. 6.1, overall, a visual inspection reveals that the betas is well scattered between 0.6 and 1.7. Focusing on the size and book-to-market ratio, it is observed that in most cases, the betas of small stocks are bigger than those of big stocks, implying that there is a size effect. In addition, it is found that the betas of growth stocks (low book-to-market stocks) are higher than those of value stocks (high book-to-market stocks). However, we notice that the dispersion of beta values increases up to scale 3 and then decreases with time scales.

Clearly, from this visual inspection, considering the size and book-to-market effects are required to explain the cross-section of average stock returns. In the next sub-section, we add SMB to capture the size effect. Our hypothesis for adding SMB is that if SMB captures the size effect significantly, the dispersion of betas will decrease.

---

[5]In their study, equally weighted portfolio returns have been applied.

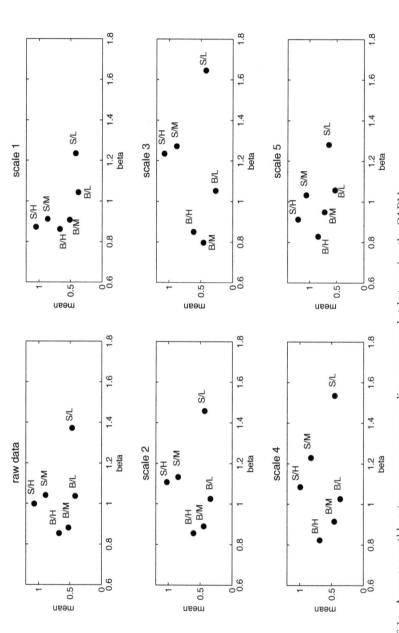

Fig. 6.1.   Average monthly return versus corresponding excess market betas using the CAPM.
Note: The wavelet scales are the following: scale 1, 2–4 month period; scale 2, 4–8 month period; scale 3, 8–16 month period; scale 4, 16–32 month period; scale 5, 32–64 month period. Note that the linear relationships are clearer with the time scales.

6.3.2. *Results using two risk factors: Excess market returns and SMB*

Following the findings of the previous sub-section, we add the SMB to capture the size effect by extending Equation (6.1).

$$R_{it}(\lambda_j) - R_{ft}(\lambda_j) = \alpha(\lambda_j) + \beta_{MKT}(\lambda_j)MKT_t(\lambda_j)$$
$$+ \beta_{SMB}(\lambda_j)SMB_t(\lambda_j) + \varepsilon_{i,t}(\lambda_j) \qquad (6.2)$$

where $SMB_t(\lambda_j)$ is the return on a portfolio of small stocks minus the return on a portfolio large stocks at scale λ_j; and $\beta_{SMB}(\lambda_j)$ is the assigned loadings on the market size (SMB) at scale λ_j.

For the size factor, proxied by the SMB, the results are presented in Table 6.3. Note that for clear presentation, we do not report the standard errors for each estimated coefficient. First, we examine the R^2s of regressions, which appear to show that the two factor model (MKT and SMB) has significant explanatory power in cross-sectional variation in average stock returns in the US. Overall, all values of R^2 increase as the time scale increases. In addition, all portfolio returns have the highest R^2 in the high time scale (scales 4 and 5), similar to the results of the CAPM model. Note that the values of R^2 in B/H do not increase, while the other portfolios' R^2 increases significantly. This result is because the SMB factor does not show significant contribution in explaining B/H portfolio.

The estimated coefficients for the SMB, $\beta_{SMB}(\lambda_j)$, is highly significant in all time scales except for B/H. Not surprisingly, the loadings on the SMB factors are positive in small stocks, but negative in big stocks. This result is expected from the construction of the SMB.

Applying the multiscaling approach to the risk factors, it is a natural question whether or not the risk factor can contribute in explaining the cross-sectional variation of average stock returns in the long-run. From Table 6.3, it is found that the SMB is an important risk factor in explaining the cross-section of average stock returns over the various time scales except for B/H. Fama and French (1997) show that a decrease in the loadings of the SMB and HML factors is generally associated with a period of unexpectedly higher cash flows. In our multiscaling approach, it is expected that a period of unexpectedly stronger cash flows in the longer time scales is less than in the shorter time scales. Thus, as the time scale increases, the loadings on the SMB is not smaller than those of the raw data. From Table 6.3, it is observed that in the longer time scale (scales 4 and 5), most portfolios show slightly higher loadings.

Table 6.3.   Estimated coefficients of the two factor (MKT and SMB) model on the wavelet domain.

		Raw data	Scale 1	Scale 2	Scale 3	Scale 4	Scale 5
S/L	α	−0.3345*	0.0012	−0.0030	0.0065	−0.0035	0.0118
	β_{MKT}	1.1467*	1.1628*	1.1262*	1.0955*	1.1442*	1.1657*
	β_{SMB}	1.0568*	1.0508*	1.0732*	1.1359*	1.0814*	0.8904*
	R^2	0.9703	0.9703	0.9736	0.9756	0.9796	0.9620
S/M	α	0.2912*	0.0001	−0.0006	−0.0094	0.0029	−0.0010
	β_{MKT}	0.8832*	0.8614*	0.9163*	0.9022*	0.9543*	0.9324*
	β_{SMB}	0.7427*	0.7210*	0.7014*	0.7631*	0.7604*	0.7708*
	R^2	0.9396	0.9291	0.9537	0.9646	0.9841	0.9734
S/H	α	0.4914*	−0.0014	0.0004	−0.0125	0.0008	0.0009
	β_{MKT}	0.8412*	0.8242*	0.8846*	0.8660*	0.8274*	0.7995*
	β_{SMB}	0.7395*	0.7028*	0.7196*	0.7604*	0.7139*	0.8619*
	R^2	0.8704	0.8674	0.8997	0.8770	0.9229	0.8808
B/L	α	−0.0244	−0.0005	0.0004	0.0033	−0.0062	−0.0021
	β_{MKT}	1.0645*	1.0509*	1.0724*	1.0993*	1.0989*	1.0692*
	β_{SMB}	−0.1276*	−0.1102*	−0.1535*	−0.0970*	−0.2021*	−0.0942*
	R^2	0.9450	0.9465	0.9496	0.9408	0.9209	0.9534
B/M	α	0.1626	−0.0001	0.0001	−0.0075	0.0068	0.0110
	β_{MKT}	0.9247*	0.9232*	0.9621*	0.8897*	0.9690*	0.9780*
	β_{SMB}	−0.2051*	−0.2321*	−0.2340*	−0.1914*	−0.1472*	−0.2253*
	R^2	0.8702	0.8876	0.8915	0.8612	0.9320	0.9684
B/H	α	0.3087*	0.0002	−0.0036	−0.0113	−0.0006	0.0024
	β_{MKT}	0.8817*	0.8744*	0.8927*	0.8754*	0.8578*	0.8491*
	β_{SMB}	−0.1283	−0.1839*	−0.1193	−0.0523	−0.0954	−0.1586
	R^2	0.7460	0.7475	0.7695	0.7645	0.8199	0.8308

Note: The table reports the estimated coefficients of the two factor (MKT and SMB) model on the wavelet domain for six size-BM portfolios. Data used are monthly US six size-BM portfolios, excess market return and SMB for the period January 1927 to December 2003. Data were obtained from the Kenneth French homepage. The heteroskedasticity adjusted error using Newey and West's (1987) method has been calculated, they are not reported due to saving the space. * indicates the significance at 5% level. The wavelet coefficients are calculated using the Daubechies least asymmetric wavelet filter of length 8 [LA(8)] up to time scale 5. The scale 1 captures oscillation with a period length 2 to 4 months. The last scale 5 captures oscillation with a period length of 32 to 64 months.

After adding the SMB in the traditional CAPM, presented in Equation (6.1), it is found that the beta $[\beta_{MKT}(\lambda_j)]$ values decreases (increases) if the values are greater (smaller) than 1 in Table 6.2. This phenomenon is clearly presented in Fig. 6.2. A visual inspection reveals that the betas is well scattered between 0.8 and 1.2. Focusing on the size and book-to-market

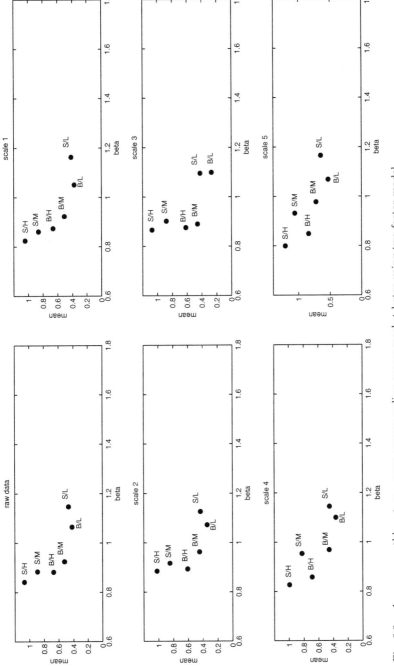

Fig. 6.2. Average monthly return versus corresponding excess market betas using two factor model.
Note: The wavelet scales are the following: scale 1, 2–4 month period; scale 2, 4–8 month period; scale 3, 8–16 month period; scale 4, 16–32 month period; scale 5, 32–64 month period. Note that the dispersion decreases compared to Fig. 6.1.

ratio, the patterns of the six size-BM portfolios are similar to Fig. 6.1, while the slope is steeper. As can be seen in Fig. 6.2, the beta values of the portfolios are close to 1, so that the dispersion of the betas decreases, compared to Fig. 6.1. Intuitively, if portfolios are correctly priced on average using the MKT, the unconditional loadings on the SMB will equal to 0.[6] Let us consider the following two equations, assuming $\text{cov}(MKT_t, SMB_t) = 0$:

$$E(R_{it}(\lambda_j) - R_{ft}(\lambda_j)) = \alpha(\lambda_j) + \beta_{MKT}(\lambda_j)E(MKT_t(\lambda_j)) \qquad (6.3)$$

$$E(R_{it}(\lambda_j) - R_{ft}(\lambda_j)) = \alpha(\lambda_j) + \beta_{MKT}(\lambda_j)E(MKT_t(\lambda_j))$$
$$+ \beta_{SMB}(\lambda_j)E(SMB_t(\lambda_j)) \qquad (6.4)$$

From Equations (6.3) and (6.4), if the unconditional loadings on the SMB is 0, two equations are identical. However, as shown in Table 6.3, our results strongly reject that the unconditional loadings on the SMB is 0 except for B/H. Given the unconditional mean return of individual portfolios, the MKT and the SMB, if $\beta_{SMB}(\lambda_j)$ is positive (negative), the loadings on the MKT will decrease (increase) to hold two equations equal. In other words, if the SMB plays a role in explaining the cross section of average stock returns, the dispersion of the market betas decreases and the market betas will be concentrated around $\beta_{MKT}(\lambda_j) = 1$, consistent with Fig. 6.2.

From Table 6.3 and Fig. 6.2, it is concluded that the SMB plays an important role in explaining the cross sectional variation of average stock returns regardless of time scales, except B/H. Overall, this result is consistent with Fama and French (1995), who report that the size factor is especially important in small stock returns.

6.3.3. *Results using three factors: Excess market returns, SMB and HML*

The purpose of this sub-section is to examine whether the model including the HML factor describes the cross section of returns in the model. We investigate the role of the HML by extending Equation (6.2) as follows:

$$R_{it}(\lambda_j) - R_{ft}(\lambda_j) = \alpha(\lambda_j) + \beta_{MKT}(\lambda_j)MKT_t(\lambda_j) + \beta_{SMB}(\lambda_j)SMB_t(\lambda_j)$$
$$+ \beta_{HML}(\lambda_j)HML_t(\lambda_j) + \varepsilon_{i,t}(\lambda_j) \qquad (6.5)$$

---

[6]Lewellen (1999) proves that the unconditional loadings on the HML are zero if assets are correctly priced on average. See Lewellen (1999:39).

where $HML_t(\lambda_j)$ is the return on a portfolio of stocks with high book-to-market ratios minus the return on a portfolio of stocks with low book-to-market ratios at scale λ_j; and $\beta_{HML}(\lambda_j)$ is the assigned loadings on the book-to-market factor (HML) at scale λ_j.

The results of three factor model are presented in Table 6.4. Note that for clear presentation, the standard errors for each estimated coefficient is not presented. The estimated coefficients for the HML, $\beta_{HML}(\lambda_j)$, is highly significant at all time scales in six size-BM portfolios. As found in Lettau and Wachter (2007), the loadings on the HML factors increase with the book-to-market ratio: growth portfolio (low BM portfolio) has negative loadings on the HML factors, while value portfolio (high BM portfolio) has positive loadings.

As mention in Section 3.2, according to Fama and French (1997), a decrease in the loadings of the SMB and HML factors is generally associated with a period of unexpectedly stronger cash flows. As in the SMB, it is expected that as the time scale increases, the loadings on the HML is not smaller than those of the raw data. This is because a period of unexpectedly stronger cash flows in the longer time scales is less than in the shorter time scales. From Table 6.3, it is observed that in the longer time scale (scales 4 and 5), most portfolios show slightly lower loadings.

According to Lakonishok *et al.* (1994), Daniel *et al.* (1998) and Cooper *et al.* (2004), the mispricing is considered as a short-term phenomenon. If the mispricing is correct, it is expected that the loadings on the HML will be significant in the short scales, while insignificant in the long scales. As can be seen in Table 6.4, all loading on the HML is significant regardless of time scales and portfolios. This implies that the idea that temporary mispricing should not explain unconditional deviations from the CAPM (Lewellen, 1999). As shown in Tables 6.3 and 6.4, six size-BM portfolios have large unconditional loadings on both the SMB and the HML. In the sense of Lewellen (1999), this result suggests that the factors do not simply capture mispricing in returns.

Moving to the values of the market betas, from Table 6.4, it is observed that the values decrease (increase) if they are greater (smaller) than 1, compared to Table 6.3. In other words, the values of the market betas are concentrating on $\beta_{MKT}(\lambda_j) = 1$. This result is clearly observed in Fig. 6.3. As mentioned in Section 3.2, if portfolios are correctly priced on average using the MKT, the unconditional loadings on the HML will equal to zero. Analogous from Equations (6.3) and (6.4), given the unconditional mean return of individual portfolios, the MKT, the SMB and the HML,

Table 6.4.   Estimated coefficients of the three factor model on the wavelet domain.

		Raw data	Scale 1	Scale 2	Scale 3	Scale 4	Scale 5
S/L	constant	−0.1850*	0.0009	−0.0031	0.0009	−0.0019	0.0104
	β_{MKT}	1.0837*	1.1094*	1.0675*	1.0196*	1.0556*	1.0397*
	β_{SMB}	1.0158*	1.0070*	1.0287*	1.0804*	1.0400*	0.8703*
	β_{HML}	−0.2579*	−0.2076*	−0.2788*	−0.3349*	−0.3177*	−0.4297*
	R^2	0.9796	0.9765	0.9825	0.9902	0.9902	0.9929
S/M	constant	0.0661	0.0005	−0.0004	−0.0044	0.0019	−0.0002
	β_{MKT}	0.9780*	0.9655*	0.9954*	0.9709*	1.0144*	1.0111*
	β_{SMB}	0.8045*	0.8063*	0.7613*	0.8133*	0.7885*	0.7834*
	β_{HML}	0.3884*	0.4044*	0.3753*	0.3029*	0.2157*	0.2684*
	R^2	0.9763	0.9722	0.9815	0.9850	0.9919	0.9919
S/H	constant	0.0888*	−0.0007	0.0008	−0.0004	−0.0020	0.0029
	β_{MKT}	1.0108*	0.9950*	1.0290*	1.0293*	0.9878*	0.9851*
	β_{SMB}	0.8500*	0.8426*	0.8291*	0.8798*	0.7890*	0.8915*
	β_{HML}	0.6947*	0.6635*	0.6861*	0.7205*	0.5752*	0.6332*
	R^2	0.9870	0.9847	0.9902	0.9880	0.9896	0.9937
B/L	constant	0.1473*	−0.0008	0.0002	−0.0011	−0.0039	−0.0031
	β_{MKT}	0.9922*	0.9700*	1.0090*	1.0396*	0.9682*	0.9816*
	β_{SMB}	−0.1747*	−0.1764*	−0.2016*	−0.1407*	−0.2632*	−0.1082*
	β_{HML}	−0.2963*	−0.3141*	−0.3009*	−0.2636*	−0.4687*	−0.2991*
	R^2	0.9718	0.9730	0.9748	0.9651	0.9750	0.9773
B/M	constant	−0.0520	0.0003	0.0003	−0.0030	0.0053	0.0115
	β_{MKT}	1.0151*	1.0159*	1.0356*	0.9509*	1.0538*	1.0172*
	β_{SMB}	−0.1462*	−0.1563*	−0.1783*	−0.1466*	−0.1075*	−0.2190*
	β_{HML}	0.3703*	0.3600*	0.3493*	0.2703*	0.3043*	0.1339*
	R^2	0.9226	0.9295	0.9329	0.9014	0.9610	0.9744
B/H	constant	−0.1266*	0.0011	−0.0032	0.0001	−0.0037	0.0044
	β_{MKT}	1.0651*	1.0841*	1.0475*	1.0298*	1.0359*	1.0357*
	β_{SMB}	−0.0088	−0.0121	−0.0020	0.0606*	−0.0121	−0.1288*
	β_{HML}	0.7511*	0.8148*	0.7348*	0.6813*	0.6383*	0.6369*
	R^2	0.9461	0.9504	0.9448	0.9681	0.9596	0.9844

Note: The table reports the estimated coefficients of the three factor (MKT, SMB and HML) model on the wavelet domain for six size-BM portfolios. Data used are monthly US six size-BM portfolios, excess market return, SMB and HML for the period January 1927 to December 2003. Data were obtained from the Kenneth French homepage. The heteroskedasticity adjusted error using Newey and West's (1987) method has been calculated, they are not reported due to saving the space. * indicates the significance at 5% level. The wavelet coefficients are calculated using the Daubechies least asymmetric wavelet filter of length 8 [LA(8)] up to time scale 5. The scale 1 captures oscillation with a period length 2 to 4 months. The last scale 5 captures oscillation with a period length of 32 to 64 months.

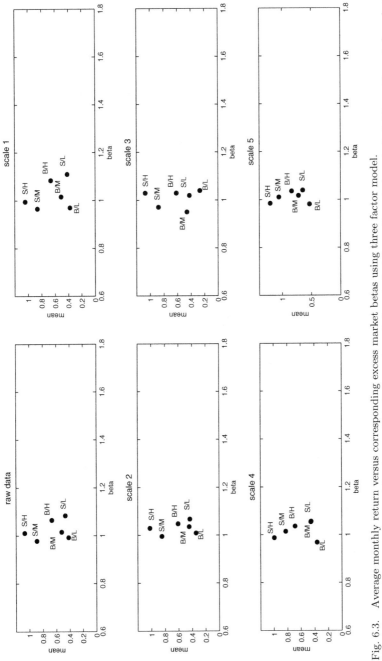

Fig. 6.3.   Average monthly return versus corresponding excess market betas using three factor model.
Note: The wavelet scales are the following: scale 1, 2–4 month period; scale 2, 4–8 month period; scale 3, 8–16 month period; scale 4, 16–32 month period; scale 5, 32–64 month period. Note that the dispersion concentrates relatively on 1, compared to Fig. 6.2.

if $\beta_{HML}(\lambda_j)$ is positive (negative), the loadings on the MKT will decrease (increase) to hold two equations equal. In other words, if the HML plays a role in explaining the cross section of average stock returns, the dispersion of the market betas decreases relatively, compared Fig. 6.2 and the market betas will be more concentrated around $\beta_{MKT}(\lambda_j) = 1$. The results are presented in Fig. 6.3. A visual inspection reveals that the betas is well scattered between 0.9 and 1.1. Focusing on the size and book-to-market ratio, the patterns of the small portfolios are similar to Fig. 6.2 showing the portfolio with high book-to-market ratio has the highest mean and the lowest beta, while the portfolio with low book-to-market ratio has the lowest mean and the highest beta. Contrast to Fig. 6.2, the big portfolios, indicated as B/L, B/M and B/H, shows a inverse relationship compared to the small portfolios.

From the results of Table 6.4 and Fig. 6.3, it is found that the HML factor captures the variations of market betas over time scales. Lewellen (1999) also finds a similar finding that B/M captures significant time variation in market betas.

Overall, the market and the book-to-market factors, proxied by excess market return and HML, play an important role in explaining the cross-sectional variation of average stock returns over all time scales. SMB can also play a role in explaining the cross-section of average returns over all time scales, except for B/H.

6.4. Concluding Remarks

This chapter uses, for the first time, the Fama–French three-factor model to explain the cross-sectional variation of the stock returns over various time scales using a new approach: wavelet analysis. This new approach is based on a wavelet multiscaling method that decomposes a given time series on a scale-by-scale basis. To evaluate the long-run relationship between six size-BM portfolio returns and four risk factors using the Fama–French model, we have adopted regression analysis in the wavelet domain.

The empirical test initially examines the traditional CAPM model. From this examination, it is found that the excess market return seems to play a role in explaining the cross-section of average stock returns regardless of time scale, evidenced by the fact that all market coefficients, market betas are statistically significant in all wavelet time scales. However, we find that there exists a size effect. Generally, market betas of small stocks are higher than those of big stocks.

As a second step investigation, we extend the traditional CAPM model by incorporating the SMB. When the SMB is incorporated in the model, the estimated coefficients for the SMB are highly significant in all time scales except for B/H. Overall, this result is consistent with Fama and French (1995), who report that the size factor is especially important in small stock returns. We find also that the significant role of the SMB in explaining the cross section of average returns makes the dispersion of the market betas concentrated around $\beta_{MKT}(\lambda_j) = 1$.

Finally, after adding the HML factor, we examine the Fama–French three factor model over the various time scales.

According to Lakonishok *et al.* (1994), Daniel *et al.* (1998) and Cooper *et al.* (2004), the mispricing is considered as a short-term phenomenon. If the mispricing is correct, it is expected that the loadings on the HML will be significant in the short scales, while insignificant in the long scales. It is found that all loading on the HML is significant regardless of time scales and portfolios. This implies that the idea that temporary mispricing should not explain unconditional deviations from the CAPM (Lewellen, 1999). Six size-BM portfolios have large unconditional loadings on both the SMB and the HML, suggesting that the factors do not simply capture mispricing in returns in the sense of Lewellen (1999).

If portfolios are correctly priced on average using the CAPM, the unconditional loadings on the HML will equal to 0. In other words, if the HML plays a role in explaining the cross section of average stock returns, the dispersion of the market betas decreases relatively and the market betas will be more concentrated around $\beta_{MKT}(\lambda_j) = 1$. This result indicates that the HML factor captures the variations of market betas over time scales. Lewellen (1999) also finds a similar finding that B/M captures significant time variation in market betas.

Regarding the results of time series regression, we find that the excess market return and HML have an explanatory power for the cross-sectional variation of six size-BM portfolio returns, while SMB is effective in explaining the cross-sectional variation for portfolios except for large stocks. Finally, we show that the Fama–French three factor model fits well since the average excess return move together in the way predicted by four risk factors.

Chapter 7

Can the Risk Factors Explain
the Cross-Section of Average Stock
Returns in the Long Run?

This chapter uses the Fama–French three factor model with momentum factor to explain the cross-sectional variation of stock returns over various time scales using a new approach. The new approach is based on a wavelet multiscaling method that decomposes a given time series on a scale-by-scale basis. The empirical results provide that market, proxied by excess market return, plays an important role in explaining stock returns over all time scales. However, SMB and HML can also play a role over all time scales, depending on the time scale. This argument is applied to the momentum factor, denoted as MOM.

7.1. Introduction

The implication of the traditional capital asset pricing model (CAPM) is a positive relationship between a security's expected return and it relative risk (beta). This relationship plays an important role in the way academics and practitioners think about risk. However, over post-1963 period academics find the failure of the CAPM. This leads the researchers to look for other factors to explain the cross-section of stock returns. One line of interpretation argues that the failure of the CAPM to explain the cross-section of average stock returns is driven by overreaction-related mispricing (Lakonishok *et al.*, 1994; Daniel and Titman, 1997). Another explanation is that the failure results from non-diversifiable risk, which is not captured by the standard CAPM (Fama and French, 1992, 1993, 1995, 1996; Vassalou, 2003). To capture non-diversifiable risk, Fama and French (1992 among others) adopt the three-factor model, by incorporating size and book-to-market factor returns into the standard CAPM. Fama and French interpret size and book-to-market factor returns as sources of risk in the ICAPM or APT framework.

If Fama and French's three factors and momentum factor represent a fundamental risk factor for explaining the cross section of stock portfolios, as argued in Liew and Vassalou (2000) and Vassalou (2003) and Lettau and Ludvigson (2001), the three factors have a predictable power to explain the cross-sectional variation of industry returns in the long run. Intuition behind this is simple. Each industry has a different cyclical movement. Therefore, due to the different cyclical movements, the risk factors may not explain the cross-sectional variation of industry returns. However, if the risk factors are fundamental, they are expected to hold in the long-run.

To examine this, we adopt a multiscaling approach. It may be an easy way to construct the quarterly, yearly return to compounding monthly returns for examining the effect of the different horizons on industry returns. However, two problems are raised when summation of monthly data to generate long horizon data. First, it is an unreliable estimator due to a handful of independent observations generated from long-horizon return series (see In and Kim, 2006). Second and most importantly, Valkanov (2003) shows that long-horizon regressions will always produce significant results, whether or not there is a structural relation between the underlying variables. This is because in a rolling summation of series integrated of order zero, i.e., $I(0)$, the new long horizon variable behaves asymptotically as a series integrated of order one, i.e., $I(1)$. As noted in Lin and Stevenson (2001), which decompose the stock and futures price series using wavelet analysis, the decomposed price series are stationary at all time scales. From this result, we consider the wavelet multiscaling decomposition as a natural tool to overcome these two problems.

The likely importance of the multiscaling approach can be found in several previous studies. For example, Levhari and Levy (1977) show that beta estimates are biased if the analyst uses a time horizon shorter than the "true" time horizon, defined as the relevant time horizon implicit in the decision-making process of investors (Gençay *et al.*, 2005). Handa *et al.* (1989) report that if we consider different return intervals, different betas can be estimated for the same stock. The arguments of these papers are supported by the widely held view that long horizon traders will essentially focus on price fundamentals that drive overall trends, whereas short-term traders will primarily react to incoming information within a short-term horizon[1] (Connor and Rossiter, 2005). Hence, important and interesting questions arise — most fundamentally, regarding whether expected stock returns relate differently to risk factors over alternate time horizons.

The main purpose of the current chapter is to examine whether or not the risk factors can explain the cross-sectional variations of industry returns over the different time scales. In other words, as Barrow (1994) argues, if fundamentals matter in the long-run, non-fundamental risks will be ignored in the long-run. To do so, we apply a relatively new technique in finance: wavelet analysis. This approach is based on a wavelet multiscaling method that decomposes a given time series on a scale-by-scale basis.[2] In the stock market, each investor has a different investment horizon. Due to the different decision-making time scales among traders, it is expected that the "true" dynamic structure of the relationship between stock returns and risk factors will *vary* over different time scales associated with those different horizons. As Bacchetta and van Wincoop (2006) point out, the obvious explanation lies in the heterogeneity of the agents on the market and, in particular, in the heterogeneity of their expectations. Certainly, due to the heterogeneity of the market, there is no reason to believe that the risk-return linkage should be identical over different time scales. In this sense, wavelet analysis provides an ideal tool for heterogeneity of market participants due to its ability.

This chapter helps to deepen our understanding of the "true" relationship between stock returns and the three risk factors over different time scales. The results therefore should be of interest to investors of all persuasions. To help examine our research question, we add one factor at a time into the traditional CAPM. More specifically, first we examine the CAPM and investigate the difference across the expanding time scales. To

---

[1]While Connor and Rossiter (2005) present this argument for the specific case of commodity markets, it is also applicable to most financial markets, and especially stock markets.

[2]Wavelet analysis is relatively new in economics and finance, although the literature on wavelets is growing rapidly. To the best of our knowledge, applications in these fields include examination of foreign exchange data using waveform dictionaries (Ramsey and Zhang, 1997), decomposition of economic relationships of expenditure and income (Ramsey and Lampart, 1998a, 1998b), the multiscale sharpe raito (Kim and In, 2005a), the multiscale relationship between stock returns and inflation (Kim and In, 2005b), systematic risk in a capital asset pricing model (Gençay *et al.*, 2003, 2005) and the multiscale hedge ratio (In and Kim, 2006).

examine the role of SMB, we add SMB into the traditional CAPM. Finally, after analyzing the role of SMB, HML is included to produce the Fama–French three-factor model.

Our empirical results show that the market, proxied by excess market return, plays an important role in explaining stock returns over all time scales. However, SMB and HML can also play a role over all time scales, depending on the time scale. This argument is applied to the momentum factor, denoted as MOM.

The remainder of this chapter is organized as follows. Section 7.2 presents the data and basic statistics. Section 7.3 discusses the empirical results. We conclude in Section 7.4 with a summary of our results.

7.2. Data and Basic Statistics

We use monthly twelve industry portfolio returns, the excess market return (MKT), SMB, and HML, Momentum factor (MOM) for the US in the period January 1964 to December 2004, obtained from the Kenneth French homepage. More specifically, to construct the twelve industry portfolios based on its four-digit SIC code, each NYSE, AMEX and NASDAQ stock to an industry portfolio is assigned at the end of June of year t. We then compute returns from July of t to June of $t+1$. At the end of June each year, all stocks are grouped on the median market capitalization of all NYSE, AMEX and NASDAQ stocks into two groups: small (S) and big (B). Stocks are also independently sorted into three BM groups: low (L), medium (M) and high (H), where L, M, and H represent the bottom 30%, middle 40%, and top 30% of stocks, respectively. BM is the ratio of book value of equity to market value of equity of a firm for the fiscal year ending in year $t - 1$. Six size-BM portfolios (S/L, S/M, S/H, B/L, B/M, B/H) are defined as the intersections of the two size and three BM groups. The monthly value-weighted average returns of each portfolio are then computed. SMB is the difference between the average returns of the three small-stock portfolios (S/L, S/M, S/H) and three big-stock portfolios (B/L, B/M, B/H). HML is the difference in returns between the two high-BM portfolios (S/H and B/H) and two low-BM portfolios (S/L and H/L). Six value-weight portfolios have been used to construct MOM.[3] More specifically, the portfolios are

---

[3]We thank Kenneth French for providing data for twelve industry portfolios, HML, SMB, MOM and the excess market returns on his website (http://mba.tuck.dartmouth.edu/pages/faculty/ken.french/)

the intersections of 2 portfolios formed on size (market equity, ME) and 3 portfolios formed on prior $(2-12)$ return. The monthly size breakpoint is the median NYSE market equity. The monthly prior $(2-12)$ return breakpoints are the 30th and 70th NYSE percentiles. MOM is the average return on the two high prior return portfolios minus the average return on the two low prior return portfolios. MKT is constructed by the difference between the value-weight return on all NYSE, AMEX, and NASDAQ stocks (from CRSP) and the one-month Treasury bill rate. Table 7.1 presents several summary statistics for the monthly data of twelve industry portfolios, the excess market returns, SMB, HML and MOM.

As shown in Panel A in Table 7.1, all sample means are positive. The sample means range from 0.358 (Telephone and Television) to 0.646 (Healthcare) among twelve industry portfolios. From the standard deviation of twelve industry portfolios, it is observed that Business Equipment has the highest volatility (6.888), while Utilities has the lowest (4.097), implying that Business Equipment (Utilities) may be more (less) sensitive to the economic condition than the other industries. Among twelve industry portfolios, the excess market returns, SMB, HML and MOM, first-order autocorrelation of monthly data ranges from -0.032 (Energy) to 0.135 (Wholesale and Retails). The measures of skewness and kurtosis are also reported to indicate whether our data (twelve industry portfolio returns, MKT, SMB, HML and MOM) are normally distributed. The signs of skewness and kurtosis vary depending on the portfolio returns, confirming that in most cases their empirical distributions have heavy tails relative to the normal distribution. Although we do not report them here, in all cases, the Jarque-Bera statistics reject normality at any conventional level of statistical significance. The Ljung-Box statistics for $k = 5$ and $k = 10$ lags and squared term indicate that significant linear and non-linear dependencies exist.

Panel B of Table 7.1 presents the sample correlation between twelve industry portfolio returns and the four risk factors. Overall, twelve industry returns show higher correlation with MKT than the other risk factors. SMB, the difference between the return on a portfolio of small stocks and the return on a portfolio of large stocks, shows positive correlation with industry portfolios except for Utilities, while HML, the difference between the return on a portfolio of high book-to-market stocks and the return on a portfolio of low book-to-market stocks, shows negative correlation with most industry portfolios except for Utilities. Like HML, MOM also has negative correlation with most industry portfolios except for Healthcare.

Table 7.1.   Descriptive statistics.

Panel A. Basic statistics

	Mean	Std. Dev.	Skewness	Kurtosis	LB(5)	LB(10)	LB²(5)	LB²(10)	ρ
Consumer Nondurables	0.625	4.532	−0.267	2.141	10.644* (0.005)	12.554 (0.084)	20.182* (0.00)	37.006* (0.00)	0.109
Consumer Durables	0.374	5.770	−0.208	1.675	13.268* (0.001)	23.465* (0.001)	4.539 (0.103)	19.379* (0.007)	0.132
Manufacturing	0.450	5.142	−0.414	2.436	4.966 (0.084)	8.251 (0.311)	4.365 (0.113)	12.810 (0.077)	0.066
Energy	0.636	5.268	0.216	1.712	4.060 (0.131)	7.425 (0.386)	18.508* (0.00)	68.842* (0.00)	−0.032
Chemicals	0.461	4.785	−0.180	2.499	1.958 (0.376)	3.478 (0.838)	10.805* (0.005)	27.921* (0.00)	0.003
Business Equipment	0.548	6.888	−0.168	1.250	3.282 (0.194)	6.649 (0.466)	109.194* (0.00)	176.098* (0.00)	0.056
Telephone and Television	0.358	4.618	−0.040	1.444	11.675* (0.003)	16.064* (0.025)	76.684* (0.00)	120.988* (0.00)	0.030
Utilities	0.367	4.097	0.107	1.135	18.938* (0.00)	21.757* (0.003)	38.452* (0.00)	70.508* (0.00)	0.027
Wholesale and Retails	0.561	5.355	−0.284	2.511	12.695* (0.002)	19.190* (0.008)	12.814* (0.002)	19.556* (0.007)	0.135
Healthcare	0.646	5.035	0.054	2.512	3.110 (0.211)	8.244 (0.312)	21.289* (0.00)	25.878* (0.001)	−0.014
Finance	0.617	5.303	−0.229	1.521	14.002* (0.001)	22.391* (0.002)	21.876* (0.00)	26.175* (0.00)	0.113
Other	0.510	5.488	−0.385	1.525	7.342* (0.026)	13.657 (0.058)	7.304* (0.026)	9.222 (0.237)	0.101
MKT	0.453	4.464	−0.495	2.015	5.805 (0.055)	9.010 (0.252)	10.445* (0.005)	17.643* (0.014)	0.057
SMB	0.266	3.275	0.543	5.584	6.902* (0.032)	11.522 (0.117)	137.854* (0.00)	138.281* (0.00)	0.067
HML	0.438	2.966	−0.014	2.656	12.078* (0.002)	15.500* (0.002)	240.922* (0.00)	344.479* (0.00)	0.131
MOM	0.846	4.078	−0.639	5.316	3.798 (0.150)	11.085 (0.150)	68.870* (0.00)	142.615* (0.00)	−0.014

Note: This table shows the basic statistics for twelve industry portfolios, excess market return (MKT), SMB, HML and Momentum (MOM) factors. The sample period ranges from January 1964 to December 2004, obtained from Ken French's homepage. SMB (HML) is the difference between the average returns of the three small-stock (two high-BM) portfolios and three big-stock (two low-BM) portfolios. MOM is the average return on the two high prior return portfolios minus the average return on the two low prior return portfolios. * indicates significant at 5% level. LB(k) and LB²(k) denote the Ljung-Box test of significance of autocorrelations of k lags for returns and squared returns, respectively. ρ is the first order autocorrelation coefficient. Skewness and kurtosis are defined as $E[(R_t − \mu)]^3$ and $E[(R_t − \mu)]^4$, where μ is the sample mean.

Table 7.1. (*Continued*)

Panel B. Correlation matrix

	Consumer non-durables	Consumer durables	Manufacturing	Energy	Chemicals	Business equipment	Telephone and television	Utilities	Wholesale and retails	Health-care	Finance	Other
MKT	0.823	0.786	0.917	0.644	0.841	0.836	0.754	0.578	0.867	0.764	0.862	0.938
SMB	0.195	0.260	0.311	0.021	0.146	0.390	0.076	-0.008	0.315	0.092	0.177	0.304
HML	-0.214	-0.095	-0.268	-0.092	-0.229	-0.609	-0.244	0.118	-0.309	-0.453	-0.151	-0.349
MOM	-0.088	-0.238	-0.129	-0.015	-0.091	-0.068	-0.176	-0.111	-0.138	0.029	-0.148	-0.064

7.3. Empirical Results

7.3.1. *Traditional CAPM context*

This sub-section examines the multiscale relationship between stock returns and excess market returns in the wavelet domain across various time scales. In our investigation, we run the following time series regression using the wavelet coefficients for scale $\lambda_j \equiv 2^{j-1}$ where $j = 1, 2, \ldots, 5$:[4]

$$R_{it}(\lambda_j) - R_{ft}(\lambda_j) = \alpha(\lambda_j) + \beta_{MKT}(\lambda_j)MKT_t(\lambda_j) + \varepsilon_{i,t}(\lambda_j) \qquad (7.1)$$

where $R_{it}(\lambda_j)$ is the return on portfolio i in calendar month t at scale λ_j; $R_{ft}(\lambda_j)$ is the risk-free return (one-month Treasury bill) in calendar month t at scale λ_j; $MKT_t(\lambda_j)$ is the excess market returns, measured by $R_{mt}(\lambda_j)$ minus $R_{ft}(\lambda_j)$, where $R_{mt}(\lambda_j)$ is the CRSP value-weighted market index return in calendar month t at scale λ_j; intercept $\alpha(\lambda_j)$ in Equation (7.1) is the abnormal return of portfolio i at scale λ_j; and $\beta_{MKT}(\lambda_j)$ is the market beta at scale λ_j.

We report the estimated coefficients of $\alpha(\lambda_j)$, $\beta(\lambda_j)$, denoted as MKT, and R^2 in Table 7.2. The first column in Table 7.2 presents the results of the original dataset for comparison. Note that these results are not from the wavelet multiscaling transform. For twelve industry portfolios we examine, the intercepts from the CAPM regressions are significantly different from zero at 10% level for Consumer Nondurables and Healthcare. Thus, as expected, a market factor does not seem to reasonably explain these stock returns.

We now turn to the results of the regression analysis focusing on the different time scales. Considering the sample size and the length of the wavelet filter, we settle on the MODWT based on the Daubechies least asymmetric wavelet filter of length 8 [LA(8)], while our decompositions go to scale 5 (equivalent to the 32–64 month period). Generally, all values of R^2 increase as the time scale increases. In addition, all portfolio returns have the highest R^2 in the high time scale (scales 4 and 5) except Energy (at scale 1) and Finance (at scale 2). This result implies that the explanatory power of the CAPM increases with the time scale. This is a similar result to that reported

---

[4]To generate asymptotically valid standard errors, in this and all remaining regressions estimated later in this chapter, we report heteroskedasticity adjusted standard errors using Newey and West's (1987) method to insure the variance-covariance matrix is positive definite.

Table 7.2. Estimated coefficients in single risk factor: CAPM on the wavelet domain.

		Original	Scale 1	Scale 2	Scale 3	Scale 4	Scale 5
Consumer Non-durables	constant	0.247*	0.001	0.001	−0.004	0.015	0.009
		(0.134)	(0.027)	(0.040)	(0.068)	(0.068)	(0.064)
	MKT	0.836***	0.816***	0.865***	0.890***	0.821***	0.935***
		(0.045)	(0.047)	(0.055)	(0.049)	(0.053)	(0.070)
	$\bar{R}^2$	0.677	0.697	0.689	0.685	0.674	0.763
Consumer Durables	constant	−0.086	0.001	0.002	−0.009	0.001	−0.003
		(0.166)	(0.042)	(0.050)	(0.102)	(0.076)	(0.082)
	MKT	1.016***	0.929***	1.061***	1.195***	0.940***	1.180***
		(0.047)	(0.057)	(0.056)	(0.044)	(0.052)	(0.068)
	$\bar{R}^2$	0.617	0.582	0.648	0.670	0.669	0.764
Manufacturing	constant	−0.028	−0.004	0.001	−0.008	0.011	0.010
		(0.102)	(0.025)	(0.031)	(0.057)	(0.051)	(0.036)
	MKT	1.056***	1.048***	1.073***	1.151***	1.055***	1.042***
		(0.027)	(0.031)	(0.027)	(0.037)	(0.035)	(0.035)
	$\bar{R}^2$	0.840	0.838	0.867	0.851	0.855	0.925
Energy	constant	0.292	−0.003	−0.002	0.000	0.016	0.004
		(0.184)	(0.047)	(0.062)	(0.109)	(0.098)	(0.122)
	MKT	0.760***	0.876***	0.703***	0.469***	0.773***	0.867***
		(0.050)	(0.055)	(0.059)	(0.079)	(0.072)	(0.120)
	$\bar{R}^2$	0.414	0.502	0.394	0.194	0.463	0.434
Chemicals	constant	0.053	−0.003	−0.001	0.000	0.004	0.002
		(0.121)	(0.031)	(0.037)	(0.062)	(0.067)	(0.044)
	MKT	0.901***	0.942***	0.889***	0.891***	0.844***	1.028***
		(0.044)	(0.054)	(0.051)	(0.041)	(0.051)	(0.039)
	$\bar{R}^2$	0.707	0.722	0.702	0.732	0.684	0.891
Business Equipment	constant	−0.036	−0.001	0.007	0.003	−0.028	−0.014
		(0.175)	(0.046)	(0.052)	(0.085)	(0.098)	(0.072)
	MKT	1.290***	1.280***	1.234***	1.417***	1.231***	1.124***
		(0.055)	(0.065)	(0.065)	(0.056)	(0.075)	(0.069)
	$\bar{R}^2$	0.698	0.677	0.702	0.787	0.687	0.789
Telephone and Television	constant	0.005	0.000	−0.001	0.004	−0.003	−0.023
		(0.144)	(0.034)	(0.042)	(0.075)	(0.067)	(0.084)
	MKT	0.780***	0.809***	0.758***	0.716***	0.747***	0.600***
		(0.043)	(0.048)	(0.056)	(0.045)	(0.055)	(0.076)
	$\bar{R}^2$	0.567	0.582	0.574	0.551	0.626	0.440
Utilities	constant	0.127	0.002	−0.001	0.006	0.030	0.002
		(0.150)	(0.039)	(0.046)	(0.083)	(0.085)	(0.060)
	MKT	0.531***	0.544***	0.546***	0.365***	0.554***	0.777***
		(0.050)	(0.060)	(0.053)	(0.053)	(0.072)	(0.066)
	$\bar{R}^2$	0.333	0.335	0.338	0.215	0.376	0.716

(Continued)

Table 7.2.   (*Continued*)

		Original	Scale 1	Scale 2	Scale 3	Scale 4	Scale 5
Wholesale	constant	0.090	0.001	−0.002	0.005	−0.007	0.003
and		(0.130)	(0.031)	(0.044)	(0.075)	(0.061)	(0.059)
Retails	MKT	1.040***	0.969***	1.077***	1.229***	1.105***	1.112***
		(0.040)	(0.042)	(0.041)	(0.048)	(0.047)	(0.066)
	$\bar{R}^2$	0.751	0.736	0.774	0.777	0.819	0.844
Healthcare	constant	0.255*	0.001	0.002	−0.001	0.018	0.000
		(0.142)	(0.037)	(0.044)	(0.073)	(0.068)	(0.097)
	MKT	0.862***	0.894***	0.818***	0.888***	0.868***	0.832***
		(0.053)	(0.069)	(0.060)	(0.047)	(0.041)	(0.068)
	$\bar{R}^2$	0.583	0.579	0.567	0.657	0.697	0.530
Finance	constant	0.153	0.001	−0.003	−0.001	0.015	−0.010
		(0.136)	(0.031)	(0.039)	(0.069)	(0.075)	(0.075)
	MKT	1.023***	0.969***	1.097***	1.102***	1.128***	0.959***
		(0.037)	(0.042)	(0.037)	(0.048)	(0.059)	(0.061)
	$\bar{R}^2$	0.742	0.726	0.789	0.773	0.756	0.715
Other	constant	−0.012	−0.001	−0.001	0.000	0.013	0.021
		(0.087)	(0.023)	(0.030)	(0.045)	(0.044)	(0.038)
	MKT	1.153***	1.111***	1.179***	1.232***	1.273***	1.071***
		(0.022)	(0.026)	(0.026)	(0.030)	(0.030)	(0.055)
	$\bar{R}^2$	0.879	0.869	0.887	0.898	0.922	0.921

Note: This table reports the estimated coefficients in the CAPM setting on the wavelet domain for six size-BM portfolios. The heteroskedasticity adjusted errors using Newey and West's (1987) method are in parentheses. *, **, and *** indicate significant at 10%, 5% and 1% levels. The wavelet coefficients are calculated using the Daubechies least asymmetric wavelet filter of length 8, up to time scale 5. The scale 1 captures oscillation with a period length 2 to 4 months. The last scale 5 captures oscillation with a period length of 32 to 64 months.

by Gençay *et al.* (2003),[5] who conclude that the predictions of the CAPM are more relevant in the medium-long run as compared to short horizons.

From the results of the regression analysis on the wavelet time domain, presented in Table 7.2, two things are worth noting. First, the intercepts are not significantly different from zero across the twelve industry portfolios regardless of time scales. Considering that the scaling coefficients capture the underlying smooth behavior of the time series, while the wavelet coefficients represent the deviations from the baseline smooth behavior, the abnormal return may occur at smoothing trends of portfolio returns.

---

[5]In their study, equally weighted portfolio returns have been applied.

Second, the excess market return is important regardless of time scale, evidenced by the fact that all market coefficients, $\beta_{MKT}(\lambda_j)$, are statistically significant at 1% level in all wavelet time scales. Fig. 7.1 is plotted to examine further the relationship between the betas and average returns across the twelve industry portfolios. Fig. 7.1 illustrates average monthly returns of twelve industry portfolios (vertical axis) and corresponding betas at different wavelet scales.[6]

From Fig. 7.1, overall, a visual inspection reveals that the betas are well scattered between 0.4 and 1.4. Focusing on the dispersion, it is observed that the degree of the dispersion is highest at scale 3, equivalent to 8–16 month period dynamics. However, we notice that the dispersion of beta values increases up to scale 3 and then decreases with time scales. Clearly, from this visual inspection, examining the relationship between risk factors and portfolio returns, it is important to analyze at different time scales.

7.3.2. *Fama–French three factor model*

Following the findings of the previous sub-section, we add the SMB to capture the size effect by extending Equation (7.1).

$$R_{it}(\lambda_j) - R_{ft}(\lambda_j) = \alpha(\lambda_j) + \beta_{MKT}(\lambda_j)MKT_t(\lambda_j) + \beta_{SMB}(\lambda_j)SMB_t(\lambda_j)$$
$$+ \beta_{HML}(\lambda_j)HML_t(\lambda_j) + \varepsilon_{i,t}(\lambda_j) \tag{7.2}$$

where $SMB_t(\lambda_j)$ is return on a portfolio of small stocks minus the return on a portfolio large stocks, $HML_t(\lambda_j)$ is the return on a portfolio of stocks with high book-to-market ratios minus the return on a portfolio of stocks with low book-to-market ratios at scale λ_j; and $\beta_{SMB}(\lambda_j)$ and $\beta_{HML}(\lambda_j)$ is the "size beta" and "HML beta" at scale λ_j, respectively. These results are presented in Table 7.3.[7]

First, looking at the R^2s of each regression, we observe high values with average 7.739, indicating that this three-factor formulation has significant explanatory power in the time-series variation of US stock returns. Generally, R^2 tends to increase with the time scale. In addition, all portfolio returns have the highest R^2 in one of the two highest time scales (scales 4 and 5), similar to the results found for the CAPM. Also worthy of note is the fact that, compared to the counterpart results in Table 7.2,

---

[6]For clear presentation, we do not indicate the industry name on the figure.
[7]Note that for clear presentation, we do not report the standard errors for each estimated coefficient.

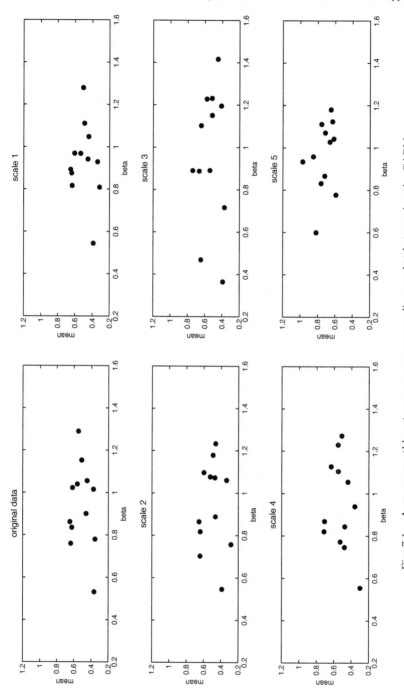

Fig. 7.1.   Average monthly return versus corresponding market betas using the CAPM.
Note: The wavelet scales are the following: scale 1, 2–4 month period; scale 2, 4–8 month period; scale 3, 8–16 month period; scale 4, 16–32 month period; scale 5, 32–64 month period.

Table 7.3. Estimated coefficients of the three-factor model on the wavelet domain.

		Original	Scale 1	Scale 2	Scale 3	Scale 4	Scale 5
Consumer	constant	0.123	0.001	0.002	−0.004	0.005	0.005
Nondurables	MKT	0.909***	0.884***	0.944***	0.955***	1.034***	0.856***
	SMB	−0.042	−0.040	−0.106*	−0.060	−0.318***	0.141
	HML	0.232***	0.237***	0.187***	0.127	0.243**	−0.200*
	$\bar{R}^2$	0.697	0.719	0.706	0.691	0.745	0.776
Consumer	constant	−0.436***	0.002	0.003	−0.002	−0.004	−0.002
Durables	MKT	1.145***	1.076***	1.166***	1.177***	1.014***	1.386***
	SMB	0.141*	0.036	0.125	0.408***	0.151	0.316***
	HML	0.579***	0.553***	0.574***	0.604***	0.343***	0.814***
	$\bar{R}^2$	0.687	0.648	0.706	0.755	0.702	0.888
Manufacturing	constant	−0.194**	−0.003	0.002	−0.004	0.007	0.007
	MKT	1.106***	1.102***	1.121***	1.133***	1.115***	1.005***
	SMB	0.103*	0.063	0.028	0.261***	0.075	0.138**
	HML	0.264***	0.218***	0.227***	0.365***	0.232***	−0.065
	$\bar{R}^2$	0.859	0.849	0.879	0.894	0.871	0.929
Energy	constant	0.150	−0.002	−0.001	0.001	0.001	0.020
	MKT	0.906***	0.997***	0.867***	0.625***	1.069***	0.963***
	SMB	−0.253***	−0.180**	−0.358***	−0.165	−0.236**	−0.972***
	HML	0.325***	0.396***	0.211**	0.269*	0.543***	−0.100
	$\bar{R}^2$	0.471	0.556	0.459	0.228	0.595	0.547
Chemicals	constant	−0.042	−0.002	0.000	0.002	−0.002	−0.003
	MKT	0.990***	1.022***	0.992***	0.938***	0.945***	0.910***
	SMB	−0.137***	−0.099*	−0.260***	0.035	0.019	0.106
	HML	0.209***	0.268***	0.095	0.217***	0.282***	−0.345***
	$\bar{R}^2$	0.731	0.750	0.734	0.749	0.715	0.925
Business	constant	0.316**	−0.004	0.005	0.001	−0.010	−0.016
Equipment	MKT	1.053***	1.044***	1.000***	1.141***	0.941***	1.038***
	SMB	0.210***	0.166**	0.280***	0.335***	−0.226***	−0.037
	HML	−0.687***	−0.824***	−0.592***	−0.400***	−0.987***	−0.298***
	$\bar{R}^2$	0.788	0.789	0.782	0.831	0.861	0.805
Telephone	constant	−0.008	0.000	0.000	0.004	−0.006	−0.017
and	MKT	0.852***	0.833***	0.839***	0.901***	0.806***	0.808***
Television	SMB	−0.220***	−0.222***	−0.093	−0.291***	−0.011	−0.125
	HML	0.086	0.025	0.211***	0.158**	0.143*	0.632***
	$\bar{R}^2$	0.593	0.604	0.594	0.592	0.635	0.597
Utilities	constant	−0.167	0.005	0.002	0.007	0.014	0.008
	MKT	0.723***	0.735***	0.760***	0.614***	0.873***	0.901***
	SMB	−0.158***	−0.054	−0.230***	−0.337***	−0.266***	−0.205*
	HML	0.567***	0.692***	0.572***	0.305***	0.573***	0.322***
	$\bar{R}^2$	0.502	0.527	0.506	0.346	0.615	0.765

(*Continued*)

138 *An Introduction to Wavelet Theory in Finance: A Wavelet Multiscale Approach*

Table 7.3. (*Continued*)

		Original	Scale 1	Scale 2	Scale 3	Scale 4	Scale 5
Wholesale	constant	−0.011	0.002	−0.002	0.008	−0.006	−0.007
and	MKT	1.052***	1.004***	1.103***	1.112***	1.071***	1.048***
Retails	SMB	0.121	0.103	−0.009	0.312***	0.195*	0.548***
	HML	0.142	0.160	0.094	0.109	0.104	0.024
	$\bar{R}^2$	0.758	0.743	0.775	0.793	0.826	0.886
Healthcare	constant	0.494***	−0.001	0.001	−0.005	0.017	−0.005
	MKT	0.831***	0.838***	0.786***	0.934***	0.913***	0.555***
	SMB	−0.287***	−0.301***	−0.273***	−0.310***	−0.383***	−0.109
	HML	−0.338***	−0.288*	−0.461***	−0.348***	−0.263**	−0.957***
	$\bar{R}^2$	0.634	0.618	0.637	0.715	0.743	0.752
Finance	constant	−0.080	0.002	−0.002	0.000	0.003	−0.014
	MKT	1.161***	1.098***	1.238***	1.286***	1.334***	1.085***
	SMB	−0.075	−0.144**	−0.122*	−0.193***	0.009	0.470***
	HML	0.435***	0.434***	0.412***	0.316***	0.550***	0.615***
	$\bar{R}^2$	0.796	0.793	0.834	0.809	0.832	0.839
Other	constant	−0.083	−0.001	0.000	0.002	0.012	0.018
	MKT	1.170***	1.132***	1.217***	1.207***	1.279***	1.062***
	SMB	0.059	0.037	−0.029	0.121*	0.161*	0.169**
	HML	0.108**	0.090	0.113*	0.114***	0.175***	0.043
	$\bar{R}^2$	0.882	0.871	0.889	0.903	0.930	0.926

Note: This table reports the estimated coefficients of the three-factor (MKT, SMB, and HML) model on the wavelet domain for six size-BM portfolios. The heteroskedasticity adjusted errors using Newey and West's (1987) method are in parentheses. *, ** and *** indicates significant at 10, 5 and 1% level. The wavelet coefficients are calculated using the Daubechies least asymmetric wavelet filter of length 8, up to time scale 5. The scale 1 captures oscillation with a period length 2 to 4 months. The last scale 5 captures oscillation with a period length of 32 to 64 months.

the values of R^2 are considerably higher in the three-factor case. Generally, adding two other risk factors (SMB and HML) into the CAPM increases its ability to explain portfolio returns.

Applying the multiscaling approach to multiple risk factors, it is a natural question whether or not the risk factor has a role to play in the long-run. The betas for MKT are highly significant for all industry portfolios regardless of the time scales. For the estimated SMB beta, $\beta_{SMB}(\lambda_j)$, in the original data set, the SMB factor is significant for most industry returns except for Consumer Nondurables, Finance and Other. When we look at loadings on the SMB at different time scales, only two industries have all significant loadings. Even though loadings are significant at original data set, most industries have insignificant loadings at intermediate and long time scales.

Next, we examine the loadings on the HML factor across industry portfolio returns. All loadings on the HML factor are significant at the original data set at 1% or 5% significance levels except for Telephone and Television, and Wholesale and Retails. As discussed in the SMB factor, the loadings on the HML factor also depend on the time scales. Interestingly, the industry, Wholesale and Retails, do not show any significant loadings over the different time scales. Overall, these results from the betas of MKT, SMB and HML imply that the MKT factor can be considered as the most important risk factor and holds at all time scales, while the role the SMB and HML factors as risk factors are dependent on the time scales.

Fama and French (1997) show that a decrease in the loadings of the SMB and HML factors is generally associated with a period of unexpectedly higher cash flows. In our multiscaling approach, it is anticipated that a period of unexpectedly stronger cash flows in the longer time scales is much less likely and of lower (relative) magnitude than in the shorter time scales. Thus, as the time scale increases, the loadings on the SMB and the HML are not smaller than those of the raw data. However, it is observed from Table 7.3 that in the longer time scales (scales 4 and 5), most portfolios show higher loadings of the SMB except for Energy and Business Equipment. The result from the HML factor is different from that of the SMB. The loadings on the HML are decreasing as the time scale increases except for Consumer Durables, Business Equipment, Telephone and Television, and Finance. From this result, it can be concluded that the HML factor is more related to the unexpected cash flows than the SMB factor.

For more examination for the role of the SMB and HML factors on the market beta, we plot the relationship between mean returns and market beta in Fig. 7.2. By augmenting the traditional CAPM setting with the SMB and HML factors, presented in Equation (7.2), it is found that the market beta estimate generally decreases (increases) if the values are greater (smaller) than unity in Table 7.2. This phenomenon is clearly presented in Fig. 7.2. A visual inspection reveals that the market betas are well scattered between 0.6 and 1.4, while the market betas of the portfolios are close to unity, so that the dispersion of the betas decreases, compared to Fig. 7.1. However, the degree of the dispersion is not much reduced because the significance of the loadings the SMB and the HML depends on time scales.

Intuitively, if portfolios are correctly priced on average using the MKT, the unconditional loadings on the SMB and the HML will equal zero.[8]

---

[8]Lewellen (1999:39) shows that the unconditional HML betas are zero if assets are correctly priced on average.

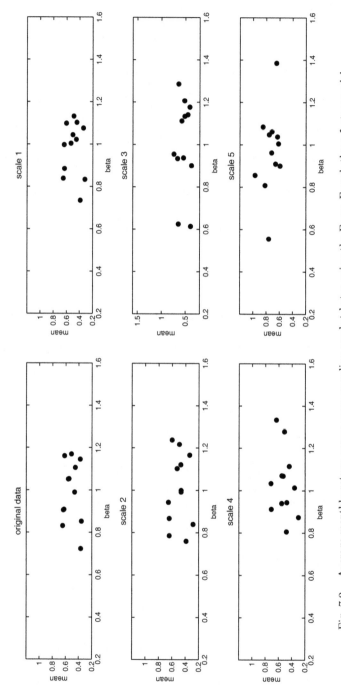

Fig. 7.2.   Average monthly return versus corresponding market betas using the Fama–French three-factor model. *Note*: The wavelet scales are the following: scale 1, 2–4 month period; scale 2, 4–8 month period; scale 3, 8–16 month period; scale 4, 16–32 month period; scale 5, 32–64 month period.

However, as shown in Table 7.3, our results reject that the unconditional loading on the SMB and HML factors is zero in the most cases. For example, given the unconditional mean return of individual portfolios, the MKT and the SMB, if $\beta_{SMB}(\lambda_j)$ is positive (negative), the loadings on the MKT will decrease (increase) to hold the expected value of Equations (7.1) and (7.2) equal. In other words, if the SMB plays a role in explaining stock returns, the dispersion of the market betas decreases and the market betas will be concentrated around $\beta_{MKT}(\lambda_j) = 1$, consistent with Fig. 7.2. In other words, the values of the market betas are "reverting" toward a value of unity. This result is clearly observed in Fig. 7.2. Based on the results presented in Table 7.3 and Fig. 7.2, it is found that the HML factor "captures" the variations of market betas over time scales. This finding is similar in nature, but more far-reaching in its scope due to the range of time scales we examine, to Lewellen (1999) who finds that BM captures significant time variation in market betas.

As such, our findings suggest that the SMB and HML factors are more likely risk factors, however their role as a risk factor depends on the time scale. This finding gives impetus to the Fama–French model research efforts that have investigated, and continue to investigate, potential explanations of the economics underlying why HML and SMB depend on the time scale and specific industry. This may be partly because of the macroeconomic risk. Chen, Roll and Ross (1986) show that industrial production is a priced macroeconomic risk variable. Therefore, if industrial production is a common factor summarizing firm-level changes of expected growth and if the role of two factors (SMB and HML) as a risk factor depends on the time scale, the movements of two risk factors (SMB and HML) and industrial production may be different from each other at specific time scales. Such targeted efforts, while beyond the scope of the current chapter, are encouraged in future research.

From Table 7.3 and Fig. 7.2, it is concluded that the SMB and the HML play important roles in explaining stock returns regardless of time scales, while being dependent on the specific time scales.

7.3.3. *Fama–French three-factor model augmented by the momentum factor*

The purpose of this sub-section is to examine whether the model including the MOM factor performs better than the other models. We investigate the role of the MOM by extending Equation (7.3) as follows:

$$R_{it}(\lambda_j) - R_{ft}(\lambda_j) = \alpha(\lambda_j) + \beta_{MKT}(\lambda_j)MKT_t(\lambda_j)$$
$$+ \beta_{SMB}(\lambda_j)SMB_t(\lambda_j) + \beta_{HML}(\lambda_j)HML_t(\lambda_j)$$
$$+ \beta_{MOM}(\lambda_j)MOM_t(\lambda_j) + \varepsilon_{i,t}(\lambda_j) \qquad (7.3)$$

where $MOM_t(\lambda_j)$ is the average return on the two high prior return portfolios minus the average return on the two low prior return portfolios at scale λ_j; and $\beta_{MOM}(\lambda_j)$ is the "MOM beta" at scale λ_j.

The results for the four-factor model are presented in Table 7.4.[9] The estimated coefficient for the MOM factor, $\beta_{MOM}(\lambda_j)$, is not significant as much as those of the SMB and the HML. More specifically, the explanatory power is increasing with the time scales. For example, at scale 1, only 3 out of 12 loadings on the MOM are significant, while at scale 5, 7 loadings are significant.

Moving to the values of the market betas, from Table 7.4, it is observed that the values of the market betas are not much different from those of Table 7.3. In case the loading of the MOM is significant, the values decrease (increase) if they are greater (smaller) than unity, compared to Table 7.3. As mentioned in Section 4.2, if portfolios are correctly priced on average using the MKT, the unconditional loadings on the MOM will equal zero. Again, given the unconditional mean return of individual portfolios, the MKT, the SMB, the HML and the MOM, if $\beta_{MOM}(\lambda_j)$ is positive (negative), the loadings on the MKT will decrease (increase) to hold expectations of the two equations equal. In other words, if the MOM plays a role in explaining stock returns, the dispersion of the market betas is expected to decrease relatively compared to Fig. 7.2, and the market betas will be more concentrated around unity. However, as shown in Fig. 7.3, the degree of the dispersion of the market betas are not significantly decreased, implying that the role of the MOM as a risk factor is very limited, compared to the SMB and the HML.

Overall, the market, proxied by excess market return, plays an important role in explaining stock returns over all time scales. However, SMB and HML can also play a role over all time scales, depending on the time scale. This argument is applied to the momentum factor, denoted as MOM.

---

[9]Note that for clear presentation, the standard errors for each estimated coefficient are not presented.

Table 7.4.   Estimated coefficients of the three-factor model augmented by momentum factor on the wavelet domain.

		Original	Scale 1	Scale 2	Scale 3	Scale 4	Scale 5
Consumer	constant	0.134	0.001	0.002	−0.005	0.004	0.006
Non-	MKT	0.908***	0.886***	0.944***	0.916***	0.993***	0.861***
durables	SMB	−0.042	−0.041	−0.106*	−0.052	−0.326***	0.116
	HML	0.229***	0.242***	0.187***	0.047	0.128	−0.206**
	MOM	−0.011	0.036	0.000	−0.187***	−0.172**	−0.066
	$\bar{R}^2$	0.696	0.719	0.705	0.712	0.759	0.777
Consumer	constant	−0.222	0.002	0.004	−0.004	−0.004	−0.001
Durables	MKT	1.117***	1.061***	1.127***	1.130***	1.018***	1.396***
	SMB	0.143**	0.043	0.125	0.417***	0.152	0.272
	HML	0.527***	0.520***	0.517***	0.509***	0.355***	0.804***
	MOM	−0.212***	−0.233***	−0.219***	−0.221***	0.018	−0.116
	$\bar{R}^2$	0.708	0.680	0.728	0.771	0.701	0.891
Manufac-	constant	−0.131	−0.003	0.002	−0.005	0.007	0.010
turing	MKT	1.098***	1.100***	1.105***	1.104***	1.072***	1.032***
	SMB	0.104**	0.064	0.028	0.26***	0.066	0.019
	HML	0.248***	0.213***	0.204***	0.307***	0.112	−0.094***
	MOM	−0.063	−0.035	−0.086**	−0.136***	−0.179***	−0.312***
	$\bar{R}^2$	0.861	0.850	0.883	0.903	0.882	0.967
Energy	constant	0.067	−0.002	−0.001	0.002	0.000	0.023
	MKT	0.917***	1.004***	0.870***	0.649***	1.011***	0.987***
	SMB	−0.254***	−0.183**	−0.358***	−0.169	−0.247**	−1.075***
	HML	0.346***	0.411***	0.216***	0.316***	0.381***	−0.124
	MOM	0.082	0.112*	0.018	0.109	−0.242***	−0.271
	$\bar{R}^2$	0.474	0.563	0.458	0.234	0.616	0.564
Chemicals	constant	−0.027	−0.002	0.000	0.001	−0.002	−0.002
	MKT	0.988***	1.026***	0.974***	0.921***	0.883***	0.919***
	SMB	−0.137***	−0.101*	−0.260***	0.038	0.007	0.066
	HML	0.205***	0.276***	0.069	0.183***	0.109	−0.354***
	MOM	−0.015	0.058	−0.101	−0.081	−0.256***	−0.104**
	$\bar{R}^2$	0.731	0.752	0.740	0.752	0.745	0.929
Business	constant	0.425***	−0.004	0.005	0.001	−0.010	−0.013
Equipment	MKT	1.039***	1.033***	0.984***	1.130***	0.973***	1.057***
	SMB	0.211***	0.171***	0.280***	0.337***	−0.219***	−0.116
	HML	−0.714***	−0.848***	−0.614***	−0.423***	−0.897***	−0.316***
	MOM	−0.108*	−0.172***	−0.084	−0.052	0.133*	−0.207*
	$\bar{R}^2$	0.792	0.800	0.784	0.831	0.865	0.816
Telephone	constant	0.125	0.000	0.000	0.003	−0.006	−0.021
and	MKT	0.835***	0.824***	0.812***	0.876***	0.822***	0.774***
Television	SMB	−0.218***	−0.218***	−0.093	−0.287***	−0.008	0.023

(*Continued*)

Table 7.4.   (*Continued*)

		Original	Scale 1	Scale 2	Scale 3	Scale 4	Scale 5
	HML	0.054	0.007	0.172**	0.109	0.190*	0.667***
	MOM	−0.131**	−0.131*	−0.150**	−0.115*	0.070	0.387***
	$\bar{R}^2$	0.605	0.617	0.611	0.601	0.637	0.679
Utilities	constant	−0.160	0.005	0.002	0.006	0.014	0.005
	MKT	0.722***	0.737***	0.756***	0.596***	0.879***	0.881***
	SMB	−0.158***	−0.054	−0.230***	−0.333***	−0.265***	−0.116
	HML	0.565***	0.695***	0.567***	0.269***	0.588***	0.343***
	MOM	−0.006	0.025	−0.020	−0.085	0.022	0.234***
	$\bar{R}^2$	0.501	0.526	0.506	0.353	0.615	0.794
Wholesale	constant	0.089	0.002	−0.002	0.005	−0.006	−0.007
and	MKT	1.039***	1.000***	1.094***	1.045***	1.011***	1.045***
Retails	SMB	0.123	0.105	−0.009	0.325***	0.183*	0.562***
	HML	0.118	0.151	0.081	−0.025	−0.063	0.027
	MOM	−0.098*	−0.067	−0.052	−0.314***	−0.248***	0.037
	$\bar{R}^2$	0.763	0.746	0.776	0.829	0.845	0.886
Healthcare	constant	0.421***	−0.001	0.001	−0.005	0.018	−0.005
	MKT	0.841***	0.846***	0.795***	0.929***	0.938***	0.552***
	SMB	−0.288***	−0.304***	−0.273***	−0.309***	−0.378***	−0.095
	HML	−0.320***	−0.273*	−0.448***	−0.358***	−0.195	−0.954***
	MOM	0.072	0.108	0.050	−0.023	0.101	0.037
	$\bar{R}^2$	0.636	0.624	0.638	0.714	0.747	0.751
Finance	constant	−0.008	0.002	−0.001	−0.001	0.003	−0.016
	MKT	1.151***	1.093***	1.227***	1.263***	1.294***	1.067***
	SMB	−0.074	−0.142**	−0.122**	−0.188***	0.002	0.546
	HML	0.417***	0.424***	0.397***	0.270***	0.440***	0.633***
	MOM	−0.072	−0.073	−0.057	−0.109*	−0.164***	0.199**
	$\bar{R}^2$	0.798	0.797	0.836	0.814	0.839	0.853
Other	constant	−0.090	−0.001	0.000	0.002	0.012	0.016
	MKT	1.171***	1.134***	1.210***	1.211***	1.285***	1.039***
	SMB	0.059	0.036	−0.028	0.120*	0.162*	0.269***
	HML	0.110***	0.093*	0.103	0.121***	0.194***	0.067
	MOM	0.006	0.023	−0.039	0.018	0.027	0.263***
	$\bar{R}^2$	0.882	0.871	0.890	0.902	0.930	0.951

Note: This table reports the estimated coefficients of the four-factor (MKT, SMB, HML and MOM) model on the wavelet domain for six size-BM portfolios. The heteroskedasticity adjusted errors using Newey and West's (1987) method are in parentheses. *, ** and *** indicates significant at 10, 5 and 1% level. The wavelet coefficients are calculated using the Daubechies least asymmetric wavelet filter of length 8, up to time scale 5. The scale 1 captures oscillation with a period length 2 to 4 months. The last scale 5 captures oscillation with a period length of 32 to 64 months.

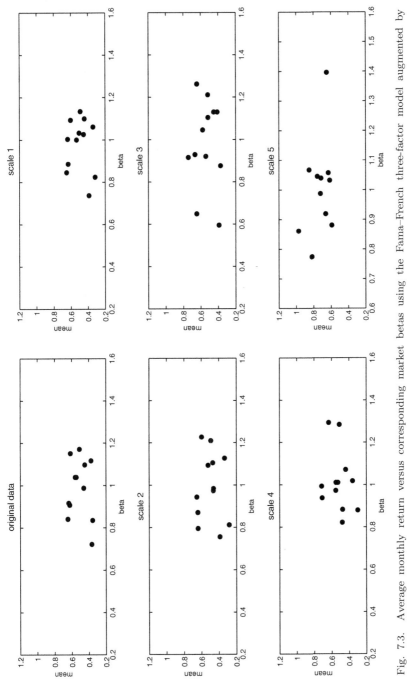

Fig. 7.3.   Average monthly return versus corresponding market betas using the Fama–French three-factor model augmented by momentum factor.
Note: The wavelet scales are the following: scale 1, 2–4 month period; scale 2, 4–8 month period; scale 3, 8–16 month period; scale 4, 16–32 month period; scale 5, 32–64 month period.

7.4. Conclusion

This chapter examines the three risk factors (excess market return, SMB, and HML) and the momentum factor to investigate whether or not they play a role as risk factors over different time scales. In this chapter, we adopt a new approach: wavelet analysis. The empirical test initially considers the traditional CAPM setting. From this examination, it is found that the excess market return is important regardless of time scale, evidenced by the fact that all market betas are statistically significant, in all time scales.

As a second step, we extend the traditional CAPM setting by incorporating the SMB factor. When the SMB and the HML are incorporated in the model, the estimated coefficients for the SMB and the HML are highly significant in specific time scales. If a decrease in the loadings of the SMB and HML factors is generally associated with a period of unexpectedly higher cash flows, as shown in Fama and French (1997), it is anticipated that a period of unexpectedly stronger cash flows in the longer time scales is much less likely and of lower (relative) magnitude than in the shorter time scales. However, it is observed that in the longer time scales (scales 4 and 5), most portfolios show higher loadings of the SMB except for Energy and Business Equipment. The result from the HML factor is different from that of the SMB. The loadings on the HML are decreasing as the time scale increases except for Consumer Durables, Business Equipment, Telephone and Television, and Finance. From this result, it can be concluded that the HML factor is more related to the unexpected cash flows than the SMB factor.

In addition, it is found that the HML factor "captures" the variations of market betas over time scales. This finding is similar in nature, but more far-reaching in its scope due to the range of time scales we examine, to Lewellen (1999) who finds that BM captures significant time variation in market betas.

Finally, we examine the role of the momentum factor by incorporating the MOM into the Fama and French three factor model. The result shows that the role of the MOM as a risk factor is very limited, compared to the SMB and the HML.

Chapter 8

Multiscale Relationships Between Stock Returns and Inflations: International Evidence

This chapter examines the Fisher hypothesis for four industrialized countries using the wavelet correlation and the multiscale hedge ratio. It is found that we find that at lower scales, the Fisher hypothesis is supportive in most countries, while at longer scales, it is only supportive in the US. The result for the US is consistent with Kim and In (2005b). From this result, it is concluded that the role of stock returns as an inflation hedge depends on the time scales and on the specific country.

8.1. Introduction

According to the most common version of the Fisher hypothesis, expected nominal asset returns should move one for one with expected inflation. Essentially, this implies that real stock returns are determined by real factors independently of the rate of inflation. However, most past empirical literature shows that stock returns are negatively correlated with inflation[1] (see Fama and Schwert, 1977; Barnes *et al.*, 1999). A negative relationship implies that investors, whose real wealth is diminished by inflation, can expect this effect to be compounded by a lower than average return on the stock market (Choudhry, 2001).

---

[1] Besides of examining the relationship between stock return and inflation, Madsen (2005) examines the Fisher hypothesis incorporating supply shock. He finds that the Fisher hypothesis cannot be rejected when supply shock variables are accommodated using 16 OECD countries.

In the long-term perspective,[2] Boudoukh and Richardson (1993), Solnik and Solnik (1997), and Schotman and Schweitzer (2000) examine the relationship between stock returns and inflation over long-horizons, and their results support the Fisher hypothesis as the horizon increases.[3] Recently, Kim and In (2005b) report the relationship between stock returns and inflation using wavelet analysis and report the interesting results that the relationship between stock returns and inflation is varying depending on specific time scale.

The main purpose of this chapter is to examine the relationship between stock returns and inflation over various time scales in four industrialized countries by extending Kim and In (2005b). We depart from the study of Kim and In (2005b) in three ways. First, although a general form of Fisher hypothesis implies the relationship between stock returns and expected inflation, Kim and In (2005b) use inflation, which is directly calculated from Consumer Price Index (CPI), not expected inflation. To examine the relationship between stock returns and expected inflation, we adopt a rolling Baysian Vector Autoregression (BVAR) to estimate expected inflation. Second, in contrast to Kim and In (2005b), the hedge ratio between stock returns and expected inflation has been adopted in our study to examine the hedge demand over various time scales. Third, considering the statistical inference and time series properties of stock returns and inflation, a nonparametric bootstrap is adopted in our study.

To investigate the multiscale relationships, we use wavelet analysis[4] to decompose a time series into various time scales. It provides a natural

---

[2]Examining the long-run relationship is important in at least two aspects. Firstly, many investors hold stocks over long holding periods. Therefore, it is important to know how stock prices move with inflation over longer horizons (Boudoukh and Richardson, 1993). Secondly, short-term noise, which might derive from agents fro portfolio rebalance or unexpected immediate consumption need reasons, may obscure the true long-run relationship between stock returns and inflation (Harrison and Zhang, 1999).

[3]Ely and Robinson (1997) examine whether stocks are a hedge against inflation in the long-run using 16 industrialized countries. In their study, it is found that where stock prices and goods prices appear to be cointegrated, the Fisher hypothesis that stocks move one-for-one with goods prices is rejected for every case.

[4]Wavelet analysis is relatively new in economics and finance, although the literature on wavelets is growing rapidly. To the best of our knowledge, applications in these fields include decomposition of economic relationships of expenditure and income (Ramsey and Lampart, 1998a and 1998b), the relationship between economic activity and the financial variables (Kim and In, 2003), systematic risk in a capital asset pricing model (Gençay *et al.*, 2003 and 2005), and multiscale hedge ratio (In and Kim, 2006).

platform on which we can investigate the multiscale hedge ratio between stock returns and inflation at different time scales without losing any information. The main advantage of wavelet analysis is its ability to decompose the data into several time scales and to handle non-stationary data, localization in time, and the resolution of the signal in terms of the time scale of analysis. Since it is likely that there are different decision-making time scales among traders, the true dynamic structure of the relationship between stock returns and inflation itself will *vary* over different time scales associated with those different horizons.

Our empirical results suggest that the relationship between stock returns and expected inflation is not constant, depending on the particular country. More specifically, from the wavelet variances, it is found that the wavelet variance decreases as the wavelet scale increases, implying that an investor with a short investment horizon has to respond to every fluctuation in the realized returns, while for an investor with a much longer horizon, the long-run risk is significantly less. From the multiscale hedge ratios, it is found that at the intermediate scales, stock returns can play a hedge against inflation in the most countries, while at the lower and longer scales, stocks cannot in most countries. From this result, it is concluded that the role of stock returns as an inflation hedge depends on the time scales and on the particular country.

This chapter is organized as follows. Section 8.2 describes how to derive the hedge ratio between stock return and expected inflation and the bootstrap approach. In Section 8.3, we present the data and the empirical results. In Section 8.4, a summary and concluding remarks are presented.

8.2. Research Methodologies

8.2.1. *The multi-scale hedge ratio*

For examining the multiscale hedge ratio, we use a single period hedge ratio, derived by Schotman and Schweitzer (2000). By extending a single period model and incorporating the results of wavelet analysis, we obtain the multiscale hedge ratio as follows:

$$HR^j = \frac{Cov_{i,\pi}(\lambda_j)}{\sigma_i^2(\lambda_j)} \tag{8.1}$$

where i and π indicate the rate of return in the country i and inflation, respectively. In this specification, HR^j indicates the wavelet multiscale hedge ratio at scale λ_j, which can be varying depending on the wavelet scales (i.e., investment horizons). This estimation is important for investors to determine whether the stock returns provide a hedge against inflation or not depending on the investment horizon. For example, if one invests in the long-run, the short-run hedge ratio will give wrong information to investors. Therefore, we need to estimate the multi-scale hedge ratio for more appropriate investment.

8.2.2. *The bootstrap approach*

The bootstrap is a nonparametric method which allows us to estimate the distribution of an estimator or test statistic by re-sampling one's data or a model estimated from the data. In this chapter, we apply the bootstrap to examine the relationship between stock returns and inflation using the hedge ratio and the correlations. The nonparametric bootstrap method is relevant for studying the relationship between stock returns and inflation for at least two reasons. First, to measure the hedge ratio and correlation, the covariance matrix is required. From this perspective, the bootstrap frees the researcher from having to estimate the entire covariance matrix characterizing the joint distribution of individual funds (Kosowski *et al.*, 2007). Specifically, the distribution of the hedge ratio and correlations depends on this covariance matrix, which is generally impossible to estimate with precision. Second, stock returns and inflation show the time series properties such as a serial correlation [see Venetis and Peel (2005) for stock returns and Pivetta and Reis (2007) for inflation]. However, through the refinement of the bootstrap, we are allowed to deal with unknown time series dependencies, resulting from heteroskedasticity or serial correlation (see Kosowski *et al.*, 2007).

More specifically, we adopt the stationary bootstrap of Politis and Romano (1994) and the algorithm of Politis and White (2004) to determine average block size for each variable. We use 5,000 bootstrap replications. Basically, the purpose of using the bootstrap is simply to find out whether or not the hedge ratio hinges on sample variability. Therefore, to obtain the distribution of the hedge ratio, we re-sample the vector of stock returns and inflation. To be more specific, at first stage, we re-sample 5,000 replications of vector of stock returns and inflation using the stationary bootstrap. At the second stage, we reproduce the same results with the original data series using wavelet multiscaling approach and using these reproduced results, we construct the confidence interval for the hedge ratio.

8.3. Data and Empirical Results

The data employed are monthly Treasury Bill rate (Call Money rate for Japan), Industrial Production, Consumer Price Index (CPI) and stock prices in 4 industrialized countries, namely, France, Japan, the UK and the US, obtained from IFS (International Financial Statistics) of the IMF (International Monetary Fund)[5]. Monthly CPI, Industrial Production and Treasury Bill rate are used to estimate the expected inflation.

Expected inflation is measured as the one month ahead forecast of a three-variable rolling Bayesian Vector Autoregression (BVAR) model, motivated by the general Phillips curve model. This approach has been adopted in the studies of Hall and Krieger (2000), and Sims (2001) to generate inflation forecast. In particular, Stock and Watson (1999) show that a generalized Phillips Curve's mod based on real aggregate activity tends to outperform many alternative models. The three variables include monthly inflation rate, Industrial Production (IP) growth rate, and T-bill rate.[6]

The purpose of this chapter is to examine the relationship between stock returns and expected inflation. In addition, we also investigate the

Table 8.1.   Correlation with inflations and expected inflations.

	France	Japan	UK	US
Inflation	−0.026*	−0.029*	0.073*	−0.008*
	(−0.025)	(−0.026)	(0.066)	(−0.006)
	[0.046]	[0.049]	[0.069]	[0.056]
Expected inflation	−0.006*	0.011*	0.066*	−0.100*
	(−0.001)	(0.009)	(0.058)	(−0.105)
	[0.053]	[0.041]	[0.067]	[0.064]

Note: Expected inflation for each country is estimated using a rolling Baysian Vector Autoression of three variables: percentage of log-difference of CPI, percentage of log-difference of Industrial Production, and Treasury Bill rate (Call Money rate for Japan), motivated by a generalized Phillips curve model. * indicates statistically significance at 5% level.

---

[5]For more detail about the data, refer to Appendix A.
[6]To generate the one month ahead forecast, the number of lag is chosen to be 12. The 3-month Treasury Bill rate has been used in a generalized Phillips curve model, except Japan. Due to unavailability of the Treasury Bill rate for Japan, we use the monthly Call Money rate. CPI is transformed by percentage of log-difference to calculate the percentage changes of inflation, Industrial production is transformed by percentage of log-difference to calculate the percentage changes of Industrial production, while Treasury Bill rate is adopted after taking the logarithm.

relationship between stock returns and inflation for comparison. Table 8.1 presents correlation with inflation and expected inflation for each country with the average correlations in the parentheses and the corresponding standard deviation in the brackets,[7] calculated from the stationary bootstrap method. As can be seen in Table 8.1, Japan shows a negative relationship with inflation, while a positive relationship with expected inflation. This table present that the results of expected inflation are supportive of Fisher Hypothesis in Japan and the UK, while those of inflation are in only the UK.

To analyze our purpose, the wavelet filters should be chosen prior to decompose our data set. Considering the sample size and the length of the wavelet filter, we settle on the Daubechies extremal phase wavelet filter of length 4 [hereafter, denoted as D(4)], while our decompositions go to level 6. Since we use monthly, scales 1 represents 2–4 month period dynamics. Equivalently, scales 2, 3, 4, 5 and 6 correspond to 4–8, 8–16, 16–32, 32–64, and 64–128 month period dynamics, respectively.

First, we examine the wavelet variances of stock returns and inflation for each country. Fig. 8.1 illustrates the logs of the wavelet variance of stock returns, inflation and expected inflation for 4 countries. The straight line indicates the wavelet variances and the dotted line indicates the 95% confidence intervals against the various time scales. Note that the variance-versus-wavelet scale curves show a broad peak at the lowest scale. This result implies in all countries that an investor with a short investment horizon has to respond to every fluctuation in the realized returns, while for an investor with a much longer horizon, the long-run risk is significantly less. The wavelet variances of inflation are very stable and close to zero over the wavelet scale for all countries. This result is generally accepted in the literature. As indicated in Schotman and Schweitzer (2000), the volatility of stock returns is usually much higher than that of inflation.

Fig. 8.1 can also be related to the well-known variance scaling. The scaling analysis has been applied to investigate long-term dependence in the financial time series. In the investor's point of view, the existence of long-term dependence implies that the investment risk is not only a function of the type of asset being considered, but also the investor's preferred investment horizon (Holton, 1992). The fact that the variances of stock

---

[7]The average correlations and the standard deviations are calculated by the 5,000 replications. After constructing the 95% confidence interval, all estimated correlations are located in the confidence interval. However, we do not report the 95% confidence interval in this table for the sake of brevity, while it is available on request.

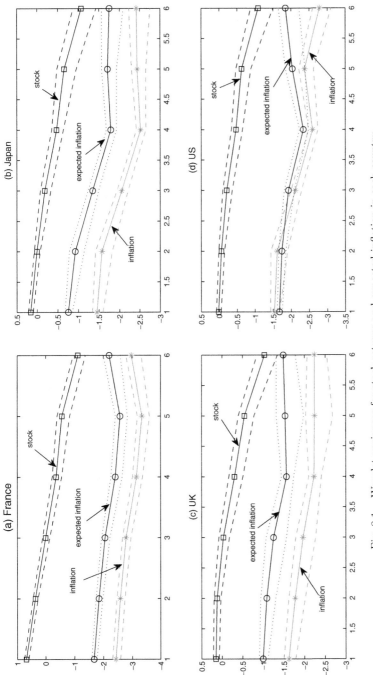

Fig. 8.1. Wavelet variances for stock returns and expected inflation in each country.

Note: The wavelet scales is defined as follows: scale 1 is associated with 2–4 month periods, scale 2 is with 4–8 month periods, scale 3 is with 8–16 month periods, scale 4 is with 16–32 month periods, scale 5 is with 32–64 month periods, and scale 6 is with 64–128 month periods. To obtain the wavelet coefficients at each scale, the Daubechies extremal phase wavelet filter of length 4 [D(4)] is applied. The method for estimating expected inflation for each country is defined as in Table 8.1. The straight lines indicate the wavelet variances, while the dotted lines are indicate 95% confidence interval. For the construction of the confidence interval, see Gençay *et al.* (2002:242).

Table 8.2.   Estimated Hurst exponents.

	Stock	Inflation	Expected Inflation
France	0.149**	0.237**	0.231**
	(6.612)	(10.210)	(10.774)
Japan	0.232**	0.191**	0.187**
	(9.005)	(5.905)	(6.635)
UK	0.286**	0.252**	0.280**
	(8.399)	(13.679)	(12.994)
US	0.274**	0.201**	0.301**
	(9.586)	(5.571)	(14.720)

Note: The Hurst exponents are estimated using Equation (1.66).
The *t*-values are in the parentheses. Expected inflation for each
country is estimated using a rolling Baysian Vector Autoression
of three variables: percentage of log-difference of CPI, percentage
of log-difference of Industrial Production, and Treasury Bill rate
(Call Money rate for Japan), motivated by a generalized Phillips
curve model. ** indicates statistically significance at 5% level.

returns are decreasing with time scales could mean that the series are anti-
correlated. To examine this fact, the Hurst exponent H is estimated to
identify the relative stability and long-term dependence for stock returns,
inflation and expected inflation using Equation (1.66). The estimated Hurst
exponents are reported in Table 8.2. The results indicate that all data series
for four countries are anticorrelated (antipersistent) with 1% significance
level. In other words, the results suggest that stock returns and inflation
follow a mean-reverting process, which has some degrees of memory.

In addition to examining the variances, it is natural to ask how stock
returns are associated with inflation. Note that our purpose in this chapter
is to examine the relationship between stock returns and expected inflation
(inflation) in four industrialized countries. It is of interest to examine
whether the correlations between stock returns and expected inflation
(inflation) change as time horizon increases. Fig. 8.2 shows the wavelet
correlations of the stock returns with expected inflation and inflation
against the wavelet scales. As can be seen in Fig. 8.2, the relationship
depends on a specific country. For France, the correlation with expected
inflation is positive up to scale 4 (equivalent to 16–32 month periods), while
the correlation with inflation is negative at scales 1, 5 and 6. The UK shows
that correlation with expected inflation and inflation is positive except scale
6 (equivalent to 64–128 month periods). However, the correlation in US is
always negative at all time scales except for the correlation between

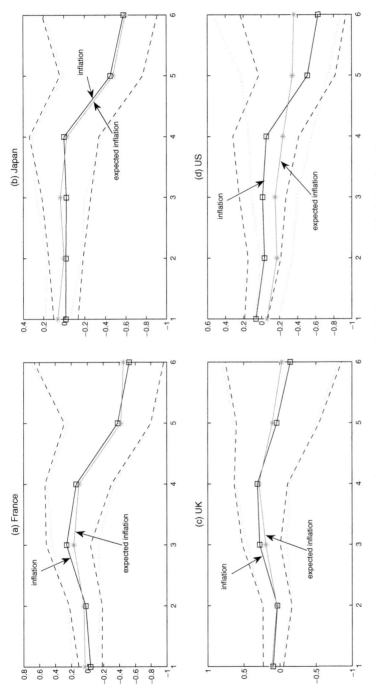

Fig. 8.2. Wavelet correlations of stock returns with inflation and expected inflation in each country.

Note: The wavelet scales is defined as follows: scale 1 is associated with 2–4 month periods, scale 2 is with 4–8 month periods, scale 3 is with 8–16 month periods, scale 4 is with 16–32 month periods, scale 5 is with 32–64 month periods, and scale 6 is with 64–128 month periods. To obtain the wavelet coefficients at each scale, the Daubechies extremal phase wavelet filter of length 4 [D(4)] is applied. The method for estimating expected inflation for each country is defined as in Table 8.1. The straight lines indicate the wavelet correlations, while the dotted lines are indicate 95% confidence interval.

stock returns and inflation at scale 1. Overall, in the most countries except the US, the positive correlation with expected inflation is observed up to the intermediate scale (scales 3 and 4), while the negative correlation at the longer scale. In Japan and the US, the correlation with inflation is negative at most time scales. This result implies that the correlation between stock return and inflation depends on the time scales and on the country.

Finally, we examine the hedge ratio between stock returns and inflation in 4 industrialized countries. When we use wavelet analysis, it is natural to ask whether the short-run hedge ratio is similar to that of long-run. Schotman and Schweitzer (2000) find that the hedge ratio is horizon sensitive and inflation plays a crucial role in the hedge ratio. Moving to our main focus on this chapter that stock returns provides the hedge against inflation at different time scales, the results, calculated using Equation (8.1), between stock returns and inflations, and between stock returns and expected inflations are presented in Tables 8.2 and 8.3. In Tables 8.2 and 8.3, we also report the average hedge ratios and the corresponding standard deviations, which are calculated from 5,000 replications using the stationary bootstrap. As can be seen in Tables 8.2 and 8.3, all estimated hedge ratios are statistically significant at 5% level except for Japan at scale 1 in Table 8.2 and the US at the raw data and Japan at scale 2 in Table 8.3.

First, we briefly examine the hedge ratios between stock returns and inflations. We present the hedge ratio using the raw data for comparison with the wavelet hedge ratios. In the raw data, two out of four countries have a negative hedge ratio; while at scales 1, three out of four countries have a positive hedge ratio. At the intermediate scales (scales 3 and 4), most countries show a positive hedge ratio except France (at scale 4).

The results with expected inflation are presented in Table 8.3. Using the raw data, one out of four countries has a statistically significantly positive hedge ratio, implying that stock returns can play a role as an inflation hedge in the UK. However, at lower scales (scales 1 and 2), many countries show a positive hedge ratio except for the US. At the intermediate scales (scales 3 and 4), France and the UK have a significantly positive hedge ratio, while Japan and the US show the negative hedge ratio. However, at the longest time scale (scale 6), only the US has the positive value, while the other countries have the negative values. These results imply that at lower scales, the Fisher hypothesis is supportive in most countries, while at longer scales, it is only supportive in the US. The result for the US is consistent with Kim and In (2005b). From this result, it is concluded that

Table 8.3.   Hedge ratios with inflations.

	France	Japan	UK	US
Raw data	−0.001*	−0.009*	0.002*	0.004*
	(−0.001)	(−0.004)	(0.009)	(−0.002)
	[0.003]	[0.008]	[0.009]	[0.009]
Scale1	−0.002*	0.023	0.015*	0.011*
	(−0.002)	(−0.001)	(0.012)	(0.010)
	[0.002]	[0.009]	[0.008]	[0.009]
Scale2	0.001*	0.006*	0.008*	−0.007*
	(−0.000)	(−0.003)	(0.005)	(−0.005)
	[0.002]	[0.009]	[0.010]	[0.011]
Scale3	0.007*	0.014*	0.028*	0.005*
	(0.006)	(−0.004)	(0.021)	(−0.006)
	[0.006]	[0.010]	[0.013]	[0.011]
Scale4	−0.001*	0.014*	0.038*	0.001*
	(−0.003)	(−0.004)	(0.019)	(−0.009)
	[0.014]	[0.015]	[0.031]	[0.016]
Scale5	0.014*	−0.035*	0.029*	−0.017*
	(−0.015)	(−0.032)	(−0.005)	(−0.033)
	[0.030]	[0.036]	[0.051]	[0.041]
Scale6	0.047*	0.003*	0.071*	−0.052*
	(−0.022)	(−0.027)	(−0.039)	(−0.020)
	[0.080]	[0.068]	[0.115]	[0.081]

Note: The raw data is reported for comparison. The wavelet scales is defined as follows: scale 1 is associated with 2–4 month periods, scale 2 is with 4–8 month periods, scale 3 is with 8–16 month periods, scale 4 is with 16–32 month periods, scale 5 is with 32–64 month periods, and scale 6 is with 64–128 month periods. To obtain the wavelet coefficients at each scale, the Daubechies extremal phase wavelet filter of length 4 [D(4)] is applied. * indicates statistically significance at 5% level.

the role of stock returns as an inflation hedge depends on the time scales and on the specific country.

8.4. Summary and Concluding Remarks

In this chapter, we examine the relationship between stock returns and inflation using four industrialized countries by extending the study of Kim and In (2005b) in three ways. Firstly, following the general form of Fisher hypothesis, we use expected inflation, which is estimated a rolling three-variable BVAR, instead of *ex post* inflation. Secondly, we apply the

Table 8.4.   Hedge ratios with expected inflations.

	France	Japan	UK	US
Raw data	-0.002^*	-0.001^*	0.028^*	-0.043
	(0.000)	(0.003)	(0.018)	(-0.019)
	$[0.007]$	$[0.014]$	$[0.020]$	$[0.012]$
Scale1	0.005^*	0.029^*	0.026^*	$-0.006*$
	(0.003)	(0.019)	(0.027)	(-0.011)
	$[0.004]$	$[0.017]$	$[0.017]$	$[0.008]$
Scale2	0.005^*	-0.036	0.003^*	$-0.025*$
	(0.003)	(0.001)	(0.014)	(-0.026)
	$[0.006]$	$[0.016]$	$[0.016]$	$[0.010]$
Scale3	0.026^*	-0.023^*	0.060^*	$-0.025*$
	(0.008)	(0.003)	(0.032)	(-0.021)
	$[0.013]$	$[0.021]$	$[0.030]$	$[0.015]$
Scale4	0.002^*	-0.086^*	0.109^*	$-0.028*$
	(-0.010)	(-0.014)	(0.034)	(-0.024)
	$[0.032]$	$[0.032]$	$[0.070]$	$[0.022]$
Scale5	-0.077^*	-0.079^*	0.055^*	$-0.061*$
	(-0.036)	(-0.077)	(-0.006)	(-0.036)
	$[0.068]$	$[0.083]$	$[0.121]$	$[0.053]$
Scale6	-0.210^*	-0.014^*	-0.065^*	$0.013*$
	(-0.051)	(-0.062)	(-0.077)	(-0.034)
	$[0.183]$	$[0.150]$	$[0.273]$	$[0.125]$

Note: The raw data is reported for comparison. The wavelet scales is defined as follows: scale 1 is associated with 2–4 month periods, scale 2 is with 4–8 month periods, scale 3 is with 8–16 month periods, scale 4 is with 16–32 month periods, scale 5 is with 32–64 month periods, and scale 6 is with 64–128 month periods. To obtain the wavelet coefficients at each scale, the Daubechies extremal phase wavelet filter of length 4 [D(4)] is applied. The method for estimating expected inflation for each country is defined as in Table 8.1. * indicates statistically significance at 5% level.

hedge ratio, proposed by Schotman and Schweitzer (2000), to estimate the multiscale hedge ratio between stock returns and expected inflation. Finally, we adopt the nonparametric bootstrap by considering the statistical inference and time series properties of stock returns and inflation.

Our main findings are summarized as follows:

(1) First, from the wavelet variances, it is found that the wavelet variance decreases as the wavelet scale increases. This result implies that an investor with a short investment horizon has to respond to every fluctuation in the realized returns, while for an investor with a much longer horizon, the long-run risk is significantly less. The wavelet

variances of inflation are very stable and close to zero over the wavelet scale, which is generally accepted in the literature.

(2) From the estimated Hurst exponent H, it is found that stock returns and inflation for all countries examined follow a mean-reverting process, which have some degree of memory.

(3) When we examine the movements of the wavelet correlation, it is found that the correlation between stock return and inflation depends on the time scales and on the country.

(4) From the examination the multiscale hedge ratio, we find that at lower scales, the Fisher hypothesis is supportive in most countries, while at longer scales, it is only supportive in the US. The result for the US is consistent with Kim and In (2005b). From this result, it is concluded that the role of stock returns as an inflation hedge depends on the time scales and on the specific country.

Appendix A. Data sources

This appendix contains the data sources and sample periods necessary to duplicate all the results in this paper. All data are taken from the IFS CD-ROM.

1. Series:
   - Share prices (###62...ZF...)
   - Treasury Bill rate (###60C...ZF...)
   - Call Money rate for Japan (###60B...ZF...)
   - Industrial Production (###66...IZF...)
   - Consumer Price Index (###64...ZF...)
     where the symbol "###" indicates the three digit country code.
   - Data is from the International Monetary Fund's International Financial Statistics CD-ROM published in June 2006.
   - Sample period for which observations are available for four series:
     — France (132): 1973:1–2003:5
     — Japan (158): 1960:1–2006:2
     — UK (112): 1967:1–1999:3
     — US (111): 1967:1–2006:2

Chapter 9

Mutual Fund Performance and Investment Horizon

This chapter examines the relative rankings of mutual fund performance at different investment horizons, with a particular focus on the Sharpe ratio. This study is important because the difference between the investment horizons of investors and the performance evaluation periods of mutual fund managers can generate suboptimal investment decisions for mutual fund investors. To this end, we utilize the wavelet multiscaling approach. Our empirical results show that all Sharpe ratios at multiple investment horizons display a very high rank correlation with respect to the Sharpe ratio at the 1-day investment horizon, implying that it matters little which investment horizon is used to assess the performance of mutual funds. This result is supported by five robustness tests.

9.1. Introduction

Mutual fund performance is a central concern for financial analysts and individual investors. To evaluate the performance of a portfolio, the Sharpe ratio is widely used in a mean-variance framework. As indicated by Levy (1972), the Sharpe ratio is closely related to the investment horizon. Levy (1972) derives the multiperiod Sharpe ratio under the assumption that returns are independently and identically distributed. This approach shows that the relative rankings of portfolios may differ with the investment horizon because expected return and standard deviation increase at different rates with investment horizons. In other words, due to the horizon sensitivity of the Sharpe ratio, multiperiod Sharpe ratios at different investment horizons may show different rankings of portfolios (Best, Hodges and Yoder, 2007). In addition to horizon sensitivity, Cvitanić, Lazrak and Wang (2008) argue that the Sharpe ratio creates a tension between short and long-term performance; this is known as the horizon problem. The horizon problem arises mainly because the investment horizon of investors differs from that of fund managers.

For example, suppose that the investment horizon of an investor is one year and that mutual funds report performance quarterly. Assume, for simplicity, that investors choose a mutual fund that is top-ranked as measured by quarterly performance. In this case, because of horizon sensitivity and the horizon problem, selecting a mutual fund may not be an optimal decision. If, in this case, the relative rankings of mutual funds based on a quarterly performance measure and those based on a yearly performance measure are identical, the difference between the investment horizon of investors and the performance measurement interval are irrelevant to investor decision making.

Despite the importance of multihorizon portfolio measurement, there are few papers on the multihorizon Sharpe ratio, aside from Hodges, Taylor and Yoder (1997), Kim and In (2005a), Best *et al.* (2007) and Cvitanić *et al.* (2008). Hodges *et al.* (1997), Kim and In (2005a), Best *et al.* (2007) focus on the multihorizon performance of stocks and bonds. Hodges *et al.* (1997) find that bonds outperform stocks over sufficiently long holding periods by assuming that returns are independently and identically distributed over time, while Best *et al.* (2007) show that stocks outperform bonds at all investment horizons when autocorrelation in returns is considered. Kim and In (2005a) conclude, using wavelet analysis, that performance depends on investment horizon. Cvitanić *et al.* (2008) examine the multihorizon Sharpe ratio theoretically when returns are iid and mean reverting. However, none of these papers examines the relative rankings[1] of mutual funds over different investment horizons.

The importance of examining the relative rankings of mutual funds over the various investment horizons lies in the widespread ownership of mutual funds. According to Investment Company Institute (2010), the percentage of US households owning mutual funds increased significantly over the past 30 years. For example, 43% of all US households owned mutual funds in 2009, while less than 6% owned them in 1980. In addition, an estimated

---

[1] In relation to relative rankings of the performance measures, Eling (2008) examines whether the different performance measures generate different rankings in mutual funds, while Eling and Schuhmacher (2007) examine hedge funds. Both papers find that the choice of performance measure does not influence the ranking of funds. In other words, the result of the Sharpe ratio is similar to those of the other 12 performance measures (Jensen measure, VaR, and Sortino Ratio, etc.).

87 million individual investors owned mutual funds, accounting for 84% of total mutual fund assets in 2009. These investors include many different types of people across all age and income groups and with various financial goals. Therefore, it is expected that individual mutual fund investors have differing investment horizons, due to their different patterns of consumption (Lee *et al.*, 1990).

The purpose of this chapter is to investigate whether the relative rankings of mutual funds differ over the various investment horizons; we focus especially on the Sharpe ratio. Our study contributes to the current literature by examining the relative rankings of the multihorizon performance measurements of mutual funds. To the best of our knowledge, the current chapter is the first to investigate relative rankings of mutual funds over multiple investment horizons (i.e., short and long investment horizons). Because individual investors may have different investment horizons, observed portfolio returns can be viewed as a composite of the trading behaviors of individual investors with different investment horizons. In this perspective, wavelet analysis is a natural tool to investigate multihorizon performance measures, because it allows decomposition of the unconditional variance into different time scales, measured at the highest possible frequency (Gençay, Selçuk and Whitcher, 2005), allowing us to observe which specific investment horizons are important contributors to the time series variance. In addition, we construct the local average of the portfolio returns in each investment horizon. These properties of wavelet analysis provide us with an effective means of constructing the Sharpe ratio at different investment horizons (Kim and In, 2005a). The outcomes can be applied to performance evaluation and should therefore be of interest to both investors and fund managers.

Our empirical results show that all Sharpe ratios at different investment horizons show very high rank correlation with respect to the Sharpe ratio at the 1-day investment horizon, implying that the difference between the investment horizon of investors and the performance measurement interval does not matter to investors in selecting a mutual fund. This result is confirmed by the five robustness tests: (1) examination by Lipper fund class; (2) separating live and defunct funds; (3) considering the possible outlier effect; (4) using other performance measures (Jensen measure, Treynor ratio, and Excess Returns on VaR); and (5) adopting the stationary bootstrap.

The remainder of this chapter is organized as follows. Section 9.2 discusses the research design for the Sharpe ratio at various investment horizons. The data and empirical results are discussed in Section 9.3. Section 9.4 presents our concluding remarks.

9.2. Sharpe Ratio at Different Investment Horizons

In this section, we first describe the Sharpe ratio at different investment horizons using the wavelet approach.

Given the wavelet variance and these wavelet coefficients at each scale, the Sharpe ratio[2] at various investment horizons can be estimated as follows:

$$SR_p^w(\lambda_j) = \frac{\bar{R}_p(\lambda_j) - \bar{R}_f(\lambda_j)}{\sqrt{\tilde{v}_p^2(\lambda_j)}} \qquad (9.1)$$

where $\bar{R}_p(\lambda_j)$ and $\bar{R}_f(\lambda_j)$ are the mean values of mutual fund returns and the risk-free rate at scale λ_j.[3] In this specification, SR_p^w indicates the wavelet multihorizon Sharpe ratio of mutual fund returns, which can be varying depending on the investment horizons.

9.3. Data and Empirical Results

In this section, we document the data used in our study (Section 3.1) and the empirical results are reported in Section 3.2, while Section 3.3 presents the five robustness test results.

9.3.1. *Data*

We use daily returns for 6,599 mutual funds, obtained from the CRSP mutual fund daily database. We also use fund names and investment objectives from the CRSP mutual fund annual database to select mutual funds that invest in domestic equity. Our sample ranges from September 1, 1998 to December 31, 2009.

---

[2]It is well known that the Sharpe ratio is an appropriate measurement of portfolio performance when the portfolio returns are normally distributed. However, many financial time series do not follow the normal distribution. Therefore, use of the Sharpe ratio for the performance measurement of mutual funds is inadequate. However, the empirical comparison of Eling (2008) shows that different measures result in a largely identical ranking.

[3]These mean values are calculated using the scaling coefficients, following Gençay *et al.* (2003).

To obtain our data, we apply several filters to our initial set of observations. Since we are interested in the performance of domestic mutual funds at various investment horizons, we restrict our attention to domestic mutual funds. The funds are categorized by Lipper fund classes such as Equity income fund (hereafter EI), Large-cap Core fund (LC), Large-cap Growth fund (LG), Large-cap Value fund (LV), Mid-cap Core fund (MC), Mid-cap Growth fund (MG), Mid-cap Value fund (MV), Multi-cap Core fund (MulC), Multi-cap Growth fund (MulG), Multi-cap Value fund (MulV), Small-cap Core fund (SC), Small-cap Growth fund (SG) and Small-cap Value fund (SV). However, the Lipper fund classes still include funds that are not invested in domestic equity. We eliminate these funds by searching for keywords such as "international" and "global" and also drop real estate funds, money market funds, and government securities funds. In addition, we require that the funds have more than 1,764 daily returns (equivalent to 7 years) and more than 20 million dollars of monthly net asset value. These elimination processes leave a total of 6,599 mutual funds.

Table 9.1 shows the number of funds and average life at each Lipper class. As is evident in the table, among 6,599 funds, 5,473 funds are live and 1,126 are defunct. In terms of Lipper fund classes, LC has the highest number of funds (864 funds), while EI has the lowest number of funds (196 funds). The average life of a fund is 10.4 years (2,608.2 days). In terms of Lipper class, the average life of a fund in each class ranges from 10.2 years (MV) to 10.6 years (EI and SV).

9.3.2. *Rank correlation between investment horizons*

To analyze the relative rankings using the Sharpe ratio over diverse investment horizons, we must determine which of various available wavelet filters to use. In consideration of the balance between sample size and length of the wavelet filter, we select the Daubechies wavelet filter of length 4 (D4), while decomposing our data up to scale 8, which is associated with up to a 256-day investment horizon (approximately 1-year investment horizon).

Table 9.2 reports the Spearman and Kendall rank correlation coefficients between the Sharpe ratios[4] at different investment horizons. All

---

[4]As a risk-free rate, the 1-month Treasury Bill rate is adopted. The daily data can be found on the Ken French's website. We thank Ken French for making his data available.

Table 9.1.   Number of mutual funds in each Lipper fund class.

	Live	Defunct	Sum	Mean (day)	Mean (year)
EI	173	23	196	2,665.4	10.6
LC	688	176	864	2,624.4	10.4
LG	492	167	659	2,566.6	10.2
LV	439	85	524	2,632.6	10.4
MC	327	52	379	2,613.9	10.4
MG	337	90	427	2,600.3	10.3
MV	280	33	313	2,577.2	10.2
MulC	668	103	771	2,591.0	10.3
MulG	462	122	584	2,599.4	10.3
MulV	522	84	606	2,604.2	10.3
SC	484	81	565	2,584.8	10.3
SG	309	79	388	2,588.0	10.3
SV	292	31	323	2,659.3	10.6
Total	5,473	1,126	6,599	2,608.2	10.4

Note: The funds are categorized by Lipper fund classes such as Equity income fund (denoted as EI), Large-cap Core fund (LC), Large-cap Growth fund (LG), Large-cap Value fund (LV), Mid-cap Core fund (MC), Mid-cap Growth fund (MG), Mid-cap Value fund (MV), Multi-cap Core fund (MulC), Multi-cap Growth fund (MulG), Multi-cap Value fund (MulV), Small-cap Core fund (SC), Small-cap Growth fund (SG) and Small-cap Value fund (SV). However, the Lipper fund classes still include funds that are not invested in domestic equity. We eliminate these funds by searching for keywords such as "international" and "global" and also drop real estate funds, money market funds and government securities funds. In addition, we require that the funds have more than 1764 daily returns and more than 20 million dollars of monthly net asset values (NAVs).

Table 9.2.   Rank correlation of multihorizon sharpe ratios.

	2 day	4 day	8 day	16 day	32 day	64 day	128 day	256 day
Spearman	0.9998	0.9999	0.9995	0.9994	0.9992	0.9981	0.9953	0.9916
	(0.0000)	(0.0000)	(0.0000)	(0.0000)	(0.0000)	(0.0000)	(0.0000)	(0.0000)
Kendall	0.9888	0.9912	0.9835	0.9821	0.9794	0.9682	0.9476	0.9316
	(0.0000)	(0.0000)	(0.0000)	(0.0000)	(0.0000)	(0.0000)	(0.0000)	(0.0000)

Note: This table reports the Spearman and Kendall rank correlations between the multihorizon Sharpe ratios. The raw data have been transformed by the wavelet filter (D4) up to 256 day investment horizon. The p-values are in the parentheses.

Sharpe ratios at the different investment horizons display a very high rank correlation with respect to the Sharpe ratio at the 1-day investment horizon (calculated using the original data). The Spearman rank correlations range from 0.9916 (256-day investment horizon) to 0.9999 (4-day investment horizon). On average, the Spearman rank correlation is 0.9978. We also find a high rank correlation in the Kendall rank correlations, with average value 0.9715.

To check the significance of the rank correlations, we adopt a standard version of the Hotelling–Pabst statistic,[5] as in Eling (2008) and Eling and Schuhmacher (2007), to check for all rank correlations. The null hypothesis of this Hotelling–Pabst statistic is that the two jointly observed samples are independent. Based on the Hotelling–Pabst statistic, all rank correlations between the Sharpe ratios at various investment horizons are rejected at the significance level $\alpha = 0.01$.

From this finding, we conclude that the relative rankings of the Sharpe ratio at different investment horizons are not different at all. Thus, it does not much matter which investment horizon is used to assess the performance of mutual funds.

9.3.3. *Robustness of the findings*

To examine the robustness of our results, several robustness tests are carried out: (1) we examine the rank correlations at each Lipper fund class; (2) to account for possible survivorship bias in our results, we separately consider live funds and defunct funds; (3) to consider the possible outlier effect, we eliminate between 1 to 10 of the highest (lowest) returns from the fund return series and replace them with the 11th highest (lowest) returns; (4) as discussed in Albrecht (1998), when the Sharpe ratio is used for long-term investment decisions, there is a problem of long-term security returns not taking the normal distribution, which the Sharpe ratio requires (to address this problem, we use other performance measures, such as the Jensen measure, the Treynor ratio, and Excess Return on VaR); and (5) as indicated in Sanfilippo (2003), in order to analyze the effect of investment horizon on relative performance, we consider the well-established history of returns over long holding periods. However, the lack of non-overlapping holding-period returns for long horizons requires care when estimating the parameters of their distribution due to estimation risk, as shown in Chopra

---

[5]When there are no ties, the test statistic T is calculated as follows: $T = \frac{6 \sum_{i=1}^{N} d_i^2 - (N^3 - N)}{\sqrt{(N-1)(N+1)^2 N^2}}$, where $d_i = R(x_{1,i}) - R(x_{2,i})$ and R(.) are the ranks of the two observed values, and N is the sample size. For $N > 30$, T is approximately standard-normally distributed (Cech, 2006).

and Ziemba (1993). To address estimation risk, several papers (Sanfilippo, 2003 and Hansson and Persson, 2000, among others) adopt the bootstrap approach. We adopt the stationary bootstrap method of Politis and White (2004).

Our method is as follows. First, we examine the Spearman rank correlations between the Sharpe ratios of each Lipper fund class level at various investment horizons. The results are presented in Table 9.3.

Table 9.3.   Spearman rank correlation at Lipper class levels.

	2 day	4 day	8 day	16 day	32 day	64 day	128 day	256 day
EI	0.9999	0.9999	0.9998	0.9995	0.9994	0.9984	0.9962	0.9957
	(0.0000)	(0.0000)	(0.0000)	(0.0000)	(0.0000)	(0.0000)	(0.0000)	(0.0000)
LC	0.9999	1.0000	0.9999	0.9999	0.9998	0.9997	0.9991	0.9978
	(0.0000)	(0.0000)	(0.0000)	(0.0000)	(0.0000)	(0.0000)	(0.0000)	(0.0000)
LG	0.9999	0.9999	0.9998	0.9998	0.9998	0.9996	0.9991	0.9975
	(0.0000)	(0.0000)	(0.0000)	(0.0000)	(0.0000)	(0.0000)	(0.0000)	(0.0000)
LV	0.9999	0.9999	0.9998	0.9997	0.9995	0.9990	0.9975	0.9947
	(0.0000)	(0.0000)	(0.0000)	(0.0000)	(0.0000)	(0.0000)	(0.0000)	(0.0000)
MC	0.9990	0.9993	0.9985	0.9978	0.9973	0.9905	0.9874	0.9712
	(0.0000)	(0.0000)	(0.0000)	(0.0000)	(0.0000)	(0.0000)	(0.0000)	(0.0000)
MG	0.9996	0.9997	0.9991	0.9988	0.9979	0.9959	0.9946	0.9891
	(0.0000)	(0.0000)	(0.0000)	(0.0000)	(0.0000)	(0.0000)	(0.0000)	(0.0000)
MV	0.9985	0.9993	0.9951	0.9936	0.9922	0.9872	0.9768	0.9448
	(0.0000)	(0.0000)	(0.0000)	(0.0000)	(0.0000)	(0.0000)	(0.0000)	(0.0000)
MulC	0.9999	0.9999	0.9997	0.9996	0.9995	0.9986	0.9976	0.9949
	(0.0000)	(0.0000)	(0.0000)	(0.0000)	(0.0000)	(0.0000)	(0.0000)	(0.0000)
MulG	0.9998	0.9999	0.9996	0.9994	0.9993	0.9985	0.9979	0.9964
	(0.0000)	(0.0000)	(0.0000)	(0.0000)	(0.0000)	(0.0000)	(0.0000)	(0.0000)
MulV	0.9997	0.9999	0.9995	0.9992	0.9989	0.9970	0.9928	0.9863
	(0.0000)	(0.0000)	(0.0000)	(0.0000)	(0.0000)	(0.0000)	(0.0000)	(0.0000)
SC	0.9987	0.9994	0.9976	0.9964	0.9952	0.9912	0.9755	0.9622
	(0.0000)	(0.0000)	(0.0000)	(0.0000)	(0.0000)	(0.0000)	(0.0000)	(0.0000)
SG	0.9995	0.9997	0.9991	0.9988	0.9979	0.9958	0.9928	0.9852
	(0.0000)	(0.0000)	(0.0000)	(0.0000)	(0.0000)	(0.0000)	(0.0000)	(0.0000)
SV	0.9984	0.9993	0.9973	0.9946	0.9914	0.9825	0.9541	0.9235
	(0.0000)	(0.0000)	(0.0000)	(0.0000)	(0.0000)	(0.0000)	(0.0000)	(0.0000)

Note: This table reports the Spearman rank correlation at each fund class. The raw data have been transformed by the wavelet filter (D4) up to 256 day investment horizon. EI indicates Equity Income fund according to Lipper fund classes. Similarly, LC is Large-cap Core fund, LG is Large-cap Growth fund, LV is Large-cap Value fund, MC is Mid-cap Core fund, MG is Mid-cap Growth fund, MV is Mid-cap Value fund, MuLC is Multi-cap Core fund, MulG is Multi-cap Growth fund, MulV is Multi-cap Value fund, SC is Small-cap Core fund, SG is Small-cap Growth fund and finally, SV is Small-cap Value fund. The *p*-values are in the parentheses.

Overall, the Spearman rank correlations between the Sharpe ratio at the 1-day investment horizon, and other Sharpe ratios at various investment horizons, remain very high, indicating that the results are not influenced by Lipper fund classes. In addition, this result confirms the results of Table 9.2.

Next, we report separate results for live and defunct mutual funds to address possible survivorship bias in our results. The literature has widely documented that the survivorship bias has a significant impact on the first and second moments (e.g., Brown *et al.*, 1992). Therefore, it is important to examine whether our results are affected by survivorship bias. To do so, we separate our data by live and defunct funds. The rank correlations of live and defunct funds are reported in Table 9.4. In this table, we report the Spearman and Kendall rank correlations between the Sharpe ratios at the 1-day investment horizon and the Sharpe ratios at other investment horizons, while the Hotelling–Pabst statistic is also reported to check the significance of the rank correlations. Overall, Table 9.4 shows that the rank correlations are very high regardless of whether funds are live or defunct, suggesting that our results are not affected by survivorship bias. Again, this confirms the results in Table 9.1.

Third, we re-examine our results considering the possible effect of outliers, as in Eling (2008) and Eling and Schuhmacher (2007). To do so, we eliminate the highest (lowest) to 10th highest (lowest) fund returns and replace the 11th highest (lowest) fund returns for each fund. The estimation results are reported in Table 9.5. As in Table 9.4, we also report the Spearman and Kendall rank correlations and the Hotelling–Pabst statistic. The results are not qualitatively different from those of Table 9.1, showing that our results are not influenced by outliers.

Fourth, Albrecht (1998) shows that when the Sharpe ratio is used for long-term investment decisions, long-term security returns may not follow the normal distribution, which the Sharpe ratio requires. To address this problem, we use other performance measures, such as the Jensen measure, the Treynor ratio, and Excess Return on VaR at different investment horizons. The Jensen measure, the Treynor ratio, and Excess Return on VaR^6 of fund p at each investment horizon are calculated as follows:

$$\alpha(\lambda_j) = [\bar{R}_p(\lambda_j) - \bar{R}_f(\lambda_j)] - \beta(\lambda_j)[\bar{R}_m(\lambda_j) - \bar{R}_f(\lambda_j)] \qquad (9.2)$$

---

[6]Recently, Fernandez (2005) decomposes the VaR using the wavelet decomposition method to quantify the VaR of a portfolio for different investment horizons.

Table 9.4. Rank correlations for live and defunct funds.

		2 day	4 day	8 day	16 day	32 day	64 day	128 day	256 day
Live	Spearman	0.9997	0.9998	0.9994	0.9992	0.9989	0.9973	0.9931	0.9883
		(0.0000)	(0.0000)	(0.0000)	(0.0000)	(0.0000)	(0.0000)	(0.0000)	(0.0000)
	Kendall	0.9869	0.9897	0.9810	0.9790	0.9752	0.9609	0.9356	0.9196
		(0.0000)	(0.0000)	(0.0000)	(0.0000)	(0.0000)	(0.0000)	(0.0000)	(0.0000)
	Hotelling	(0.0000)	(0.0000)	(0.0000)	(0.0000)	(0.0000)	(0.0000)	(0.0000)	(0.0000)
Defunct	Spearman	0.9997	0.9997	0.9994	0.9991	0.9987	0.9976	0.9950	0.9895
		(0.0000)	(0.0000)	(0.0000)	(0.0000)	(0.0000)	(0.0000)	(0.0000)	(0.0000)
	Kendall	0.9884	0.9377	0.9822	0.9783	0.9739	0.9654	0.9476	0.9221
		(0.0000)	(0.0000)	(0.0000)	(0.0000)	(0.0000)	(0.0000)	(0.0000)	(0.0000)
	Hotelling	(0.0000)	(0.0000)	(0.0000)	(0.0000)	(0.0000)	(0.0000)	(0.0000)	(0.0000)

Note: This table reports the Spearman and Kendall rank correlations, and also presents the p-values of Hotelling–Pabst statistics (indicated as Hotelling) among live and defunct funds. The Hotelling–Pabst statistic T is calculated as $T = \frac{6\sum_{i=1}^{N} d_i^2 - (N^3 - N)}{\sqrt{(N-1)(N+1)^2 N^2}}$, where $d_i = R(x_{1,i}) - R(x_{2,i})$ with $R(.)$ the ranks of the two observed values, and N is the sample size. For $N > 30$, T is approximately standard normally distributed (Cech, 2006). The raw data have been transformed by the wavelet filter (D4) up to 256 day investment horizon. The p-values are in the parentheses.

Table 9.5.   Rank correlations after removing highest and lowest returns.

	2 day	4 day	8 day	16 day	32 day	64 day	128 day	256 day
Spearman	0.9998	0.9999	0.9997	0.9995	0.9993	0.9982	0.9963	0.9947
	(0.0000)	(0.0000)	(0.0000)	(0.0000)	(0.0000)	(0.0000)	(0.0000)	(0.0000)
Kendall	0.9931	0.9962	0.9904	0.9871	0.9840	0.9727	0.9568	0.9479
	(0.0000)	(0.0000)	(0.0000)	(0.0000)	(0.0000)	(0.0000)	(0.0000)	(0.0000)
Hotelling	(0.0000)	(0.0000)	(0.0000)	(0.0000)	(0.0000)	(0.0000)	(0.0000)	(0.0000)

Note: This table reports the Spearman and Kendall rank correlations, and also presents the *p*-values of Hotelling–Pabst statistics (indicated as Hotelling) after removing the possible outiers. More precisely, we eliminate the highest (lowest) to 10th highest (lowest) fund returns and replace the 11th highest (lowest) fund returns at each fund. Hotelling indicates the *p*-values of the Hotelling–Pabst statistics. The raw data have been transformed by the wavelet filter (D4) up to 256 day investment horizon. The *p*-values are in the parentheses.

$$Treynor(\lambda_j) = \frac{\bar{R}_p(\lambda_j) - \bar{R}_f(\lambda_j)}{\beta(\lambda_j)} \tag{9.3}$$

$$\text{Excess return on VaR} = \frac{\bar{R}_p(\lambda_j) - \bar{R}_f(\lambda_j)}{VaR_p(\lambda_j)},$$

$$VaR_p(\lambda_j) = -(\bar{R}_p(\lambda_j) + z_\alpha \sqrt{\tilde{v}_p^2(\lambda_j)}) \tag{9.4}$$

where $\bar{R}_m(\lambda_j)$ and $\beta(\lambda_j)$ are the mean values of market returns and the market beta, respectively, at scale λ_j. z_α denotes the α-quantile of the standard normal distribution.

Table 9.6 reports the Spearman rank correlation between the Sharpe ratio and the other performance measures at each investment horizon. Similar to Eling (2008), the Spearman rank correlations among the performance measures are quite high at all investment horizons. We also calculate the Hotelling–Pabst statistic to check the significance of the rank correlations, while all rank correlations between the performance measures at various investment horizons are rejected at significance level $\alpha = 0.01$.

As a final robustness test, we adopt the nonparametric bootstrap method. This method is relevant for studying mutual funds for at least two reasons. First, by adopting the bootstrap, one avoids having to make a priori assumptions about the distribution of mutual funds. As indicated by Lo (2002), the accuracy of the Sharpe ratio estimator depends on the statistical

Table 9.6.   Spearman rank correlations with other performance measures at each investment horizon.

	1 day				2 day				4 day			
	Sharpe	Jensen	Treynor	Excess return on VaR	Sharpe	Jensen	Treynor	Excess return on VaR	Sharpe	Jensen	Treynor	Excess return on VaR
Sharpe	1.0000	0.9170	0.9973	1.0000	1.0000	0.9455	0.9921	1.0000	1.0000	0.9567	0.9893	1.0000
Jensen		1.0000	0.9140	0.9168		1.0000	0.9352	0.9453		1.0000	0.9464	0.9565
Treynor			1.0000	0.9974			1.0000	0.9924			1.0000	0.9900
Excess return on VaR				1.0000				1.0000				1.0000

	8 day				16 day				32 day			
	Sharpe	Jensen	Treynor	Excess return on VaR	Sharpe	Jensen	Treynor	Excess return on VaR	Sharpe	Jensen	Treynor	Excess return on VaR
Sharpe	1.0000	0.9680	0.9870	1.0000	1.0000	0.9686	0.9908	0.9999	1.0000	0.9682	0.9931	0.9962
Jensen		1.0000	0.9551	0.9677		1.0000	0.9576	0.9680		1.0000	0.9587	0.9635
Treynor			1.0000	0.9881			1.0000	0.9921			1.0000	0.9910
Excess return on VaR				1.0000				1.0000				1.0000

	64 day				128 day				256 day			
	Sharpe	Jensen	Treynor	Excess return on VaR	Sharpe	Jensen	Treynor	Excess return on VaR	Sharpe	Jensen	Treynor	Excess return on VaR
Sharpe	1.0000	0.9448	0.9829	0.9960	1.0000	0.9575	0.9872	0.9955	1.0000	0.9635	0.9951	0.9932
Jensen		1.0000	0.9186	0.9401		1.0000	0.9340	0.9505		1.0000	0.9557	0.9513
Treynor			1.0000	0.9827			1.0000	0.9872			1.0000	0.9905
Excess return on VaR				1.0000				1.0000				1.0000

Note: This table reports the Spearman rank correlations of the Sharpe ratios with other different performance measures (Jensen, Treynor and excess returns on VaR) at different investment horizons. The raw data have been transformed by the wavelet filter (D4) up to 256 day investment horizon.

properties of the returns.[7] This implies that statistical significance can be better evaluated using a nonparametric bootstrap. Second, as Sharpe (1994) himself acknowledges, the Sharpe ratio may not give a reliable ranking if one or more of the assets involved is correlated with the rest of the portfolio. This suggests the need for a method such as the bootstrap, which frees the researcher from having to estimate the entire covariance matrix characterizing the joint distribution of individual funds (Kosowski *et al.*, 2007).

In this chapter, we use the stationary bootstrap of Politis and Romano (1994) with the algorithm of Politis and White (2004), whose advantage is to mimic the original model by retaining the stationary property of the original series in the re-sampled pseudo-time series (Politis and Romano, 1994). More specifically, we re-sample the vector of each fund's returns and other variables (market returns and risk-free rates) simultaneously by using the stationary bootstrap method of Politis and Romano (1994) and the algorithm of Politis and White (2004). Fundamentally, the purpose of using the bootstrap is simply to determine whether the performance of each mutual fund hinges on sample variability. Therefore, to obtain the distribution of the Sharpe ratio at each investment horizon, we re-sample the vector of mutual funds, market returns, and risk-free rates. To be more specific, at the first stage of our procedure, we re-sample 1,000 replications of vectors of mutual funds, market returns, and risk-free rates using the stationary bootstrap. At the second stage, we reproduce the same results with the original data series using a wavelet multiscaling approach and we construct the standard errors for the Sharpe ratio via these reproduced results.

Table 9.7 reports the bootstrap simulation results for the Spearman and Kendall rank correlations between multihorizon Sharpe ratios. In this table, the raw rank correlations, which are calculated using the original return series, are presented in the first row in each panel. The mean rank correlations, which are calculated via 1,000 bootstrap simulations, are reported in the second row. Finally, the bootstrapped standard errors are reported in the third row.

---

[7]According to Lo (2002), the statistical properties can vary depending on investment styles of the portfolios. Therefore, Lo argues that the Sharpe ratios are more likely accurate when evaluating mutual funds rather than hedge funds, because hedge funds have more volatile investment strategies.

Table 9.7.   Bootstrap simulation results for rank correlations.

	2 day	4 day	8 day	16 day	32 day	64 day	128 day	256 day
Panel A. Spearman rank correlation								
Raw	0.9998	0.9999	0.9995	0.9994	0.9992	0.9981	0.9953	0.9916
Mean	0.9998	0.9998	0.9995	0.9993	0.9986	0.9973	0.9948	0.9901
	(0.0001)	(0.0000)	(0.0001)	(0.0001)	(0.0006)	(0.0008)	(0.0005)	(0.0015)
Panel B. Kendall rank correlation								
Raw	0.9888	0.9912	0.9835	0.9821	0.9794	0.9682	0.9476	0.9316
Mean	0.9893	0.9892	0.9824	0.9782	0.9696	0.9570	0.9403	0.9167
	(0.0005)	(0.0020)	(0.0011)	(0.0039)	(0.0098)	(0.0113)	(0.0074)	(0.0149)

Note: This table reports the Spearman and Kendall rank correlations between the multihorizon Sharpe ratios at different investment horizons. The raw data have been transformed by the wavelet filter (D4) up to 256 day investment horizon. In this table, "raw" indicates the rank correlation calculated by the original data series, while "mean" indicates the average rank correlation using 1,000 replications by the stationary bootstrap. The standard errors using the 1,000 bootstrap replications are reported in the parentheses.

As is evident from Table 9.7, the mean rank correlations (Spearman and Kendall rank correlations) are very close to the original rank correlations, while the mean rank correlations remain very high. This implies that all Sharpe ratios at different investment horizons display a very high rank correlation with respect to the Sharpe ratio at the 1-day investment horizon. More specifically, this result shows that our results are not influenced by sample variability.

In summary, we re-examine our results using the five robustness tests. All robustness test results show that the rank correlations between the Sharpe ratios at different investment horizons are very high, implying that it does not much matter which investment horizon is used to assess the performance of mutual funds.

9.4. Concluding Remarks

This chapter examines the relative rankings of mutual fund performance at different investment horizons, with a particular focus on the Sharpe ratio. This study is important because the difference between the investment horizons of investors and the performance evaluation periods of mutual fund managers can generate sub-optimal investment decision making on the part of mutual fund investors. Accordingly, we employ the wavelet multiscaling approach to evaluate the Sharpe ratios over various investment horizons.

We find that all Sharpe ratios at different investment horizons display a very high rank correlation with respect to the Sharpe ratio at the 1-day investment horizon, implying that it does not much matter which investment horizon is used to assess the performance of mutual funds.

To examine the robustness of our results, we adopt five robustness tests: (1) examining by Lipper fund classes, (2) separating live and defunct funds, (3) considering the possible outlier effect, (4) using other performance measures (Jensen measure, Treynor ratio, and Excess Returns on VaR; and (5) adopting the stationary bootstrap. All robustness test results confirm our result that the rank correlations remain quite high between all investment horizons.

Chapter 10

A New Assessment of US Mutual Fund Returns Through a Multiscaling Approach

This chapter applies a multiscaling approach to evaluate the performance of three types of US mutual fund: Institutional, Active, and Index. Since risk and value (performance) are timescale-dependent concepts, any form of measurement, such as the frequently used Sharpe ratio or Jensen's alpha, must account for any investment horizon effect. The results of this new analysis show that Institutional funds are clearly dominant over all time scales.

10.1. Introduction

Holding period (defined below) is an important factor that has a considerable bearing on the assessment of fund performance. It is notable then, that holding period has been largely overlooked in the literature, both in the traditional and the more recent studies. Accordingly, the primary aim of this chapter is to help redress this gap in the literature by, for the first time, applying multiscaling techniques to the Sharpe ratio assessment of US mutual fund performance.

The holding period that is relevant to portfolio allocation is the length of time investors hold a stock or bond (Siegel, 1998:29). In other words, an investor's investment horizon sensitivity is critical to evaluating the performance of an investment portfolio. A private (relatively uninformed) investor might be entirely uninterested in short-term portfolio performance. In contrast, institutional investors, such as pension funds, generally hold long-term investment horizons. Still other investors, such as those influenced by feedback trading notions, may be much more attuned to short-term performance. As such, it would be of considerable interest to examine long-term investment performance as investment horizon increases.

The examination of multi-horizon performance measures has at least a two-fold importance. First, multi-horizon measures would be useful tools for fund managers in their evaluation of fund performance. For example, consider an investment company with a large number of money managers and investor clients. Suppose, for simplicity, that the investment horizon of an investor is one year and that the investment company reviews the performance of money managers every quarter. The money manager will therefore focus on the three-month performance of a portfolio, while the investor concentrate on the one-year performance. Clearly in this case, investors and money managers will make decisions over different time scales. Thus, for such an investor, the money manager may not provide the best service. Secondly, from the investor's point of view, there is a clear need to have sufficient information about the performance of mutual funds over a long period before selecting a fund. Ignoring the time-scale effect, then, might bring about a biased performance measure.

This chapter aims to contribute to the literature on the study of the performance measures of mutual fund returns by employing a multiscaling approach: wavelet analysis. To the best of our knowledge, no previous study has conducted such an investigation. Wavelet analysis, due to its nonparametric nature, offers the significant advantage of not requiring any assumptions about the distribution of returns.

The Sharpe ratio[1] has been the generally adopted measurement used to evaluate the performance of mutual funds. Such use of the Sharpe ratio is founded in the studies of Pfingsten *et al.* (2004), and Pedersen and Rudholm-Alfvin (2003), who find that different performance measures result in a largely identical ranking. The Sharpe ratio is also adopted in Lowe and Davidson III (2000) to examine successions of fund managers in the closed-end mutual fund industry. In addition, a recent study by Kim and In (2005a) proposes the multiscale Sharpe ratio to evaluate the performance of a portfolio, based on wavelet multiscaling decomposition. We further adopt a nonparametric bootstrap method for more concrete statistical inference. The bootstrap method has been applied recently by Kosowski *et al.* (2006) in relation to mutual funds.

The main advantage of wavelet analysis is its ability to decompose data into multiple time scales. Consider the large number of investors who

---

[1] As in this study, Hodges *et al.* (1997) examine the multi-period Sharpe ratio using a bootstrap method. The bootstrap method is used in their study to generate the longer term returns, while the same method is adopted for generating statistical inferences, wavelet analysis, and the movements of various time scales.

trade in securities markets and who make decisions over varying time scales. Due to the presence of these differing decision-making time scales among investors, the true dynamic structure of the relationship between variables will *vary* over the various time scales associated with those differing time horizons. While economists and financial analysts have long recognized the idea of multiple time periods in decision making, they have traditionally been effectively forced to adopt the simplistic dichotomous characterization of "short-run" versus "long-run", due to the lack of analytical tools with which to decompose data into more than two time scales (In and Kim, 2006).

In addition, we examine correlation with market index returns. The examination of this correlation over various investment horizons is important in the sense that the inclusion of mutual funds in a portfolio can potentially result in better risk–return tradeoffs.

Our results show that Institutional funds are clearly dominant over all time scales. Since risk and value (performance) are timescale-dependent concepts, any form of measurement, such as the frequently used Sharpe ratio or Jensen's alpha, must account for any investment horizon effect.

The remainder of the chapter is organized as follows. Section 10.2 discusses performance measure models. The data and empirical results are discussed in Section 10.3. Section 10.4 presents the summary and concluding remarks.

10.2. Empirical Method

10.2.1. *Multiscaling approach*

Currently, most performance studies of multi-index asset pricing models use Jensen's alpha (Jensen, 1968). Its interpretation as the risk-adjusted abnormal return of a portfolio makes it flexible enough to be used in most asset pricing specifications. Kothari and Warner (2001) consider this measure exclusively, for multi-index asset pricing models in their empirical comparison of mutual fund performance measures. However, in the present study the Sharpe ratio is adopted as in Pfingsten *et al.* (2004), and Pedersen and Rudholm-Alfvin (2003). In this section, we briefly explain derivation of the Sharpe ratio in a multiscaling context.

Given the wavelet variance and these wavelet coefficients at each scale, the Sharpe ratio at various time scales can be estimated as follows

$$SR_p^w(\lambda_j) = \frac{\bar{R}_p(\lambda_j) - \bar{R}_f(\lambda_j)}{\sqrt{\sigma^2(\lambda_j)}}, \tag{10.1}$$

where $\bar{R}_p(\lambda_j)$ and $\bar{R}_f(\lambda_j)$ are the mean values of mutual fund returns and the risk-free rate at scale λ_j. These mean values are calculated using the scaling coefficients, as in Gençay *et al.* (2003). In this specification, SR_p^w indicates the wavelet multiscale Sharpe ratio of mutual fund returns, which can vary depending on the wavelet scales (i.e., investment horizons).

10.2.2. *The bootstrap approach*

The bootstrap approach is a nonparametric method which allows us to estimate the distribution of an estimator or test statistic by re-sampling data or a model estimated from that data. In general, mutual fund returns are characterized by small sample sizes, serial correlation, volatility clustering, and non-normality. These characteristics lead us to use the bootstrap method in our analysis.

In this chapter, we apply the bootstrap method to examine the performance of mutual funds using the Sharpe ratio and to investigate correlation with market returns. As indicated in Kosowski *et al.* (2007),[2] the nonparametric bootstrap method is relevant to the study of mutual funds for at least three reasons. First, by adopting the bootstrap method, one avoids having to make *a priori* assumptions about the distribution of mutual funds. As indicated in Lo (2002),[3] the accuracy of the Sharpe ratio estimator depends on the statistical properties of returns; Lo also notes that the comparison of the Sharpe ratios between two mutual funds cannot be performed naively, due to their return characteristics. This implies that statistical significance can be better evaluated using a nonparametric bootstrap method.

Second, as Sharpe (1994) himself acknowledges, the Sharpe ratio may not yield a reliable ranking if one or more of the assets involved is correlated with the rest of the portfolio. This suggests the need for a method like the bootstrap method, which frees the researcher from having to estimate the entire covariance matrix characterizing the joint distribution of individual funds (Kosowski *et al.*, 2007). Finally, the Sharpe ratio can also be influenced by the time series properties of such investment strategies as serial correlation (see Lo, 2000). However, through refinement of the bootstrap method, we are able to deal with unknown time series

---

[2]As in Kosowski *et al.* (2007), we also use the bootstrap method of Efron (1979).
[3]Lo (2002) derives the statistical distribution of the Sharpe ratio under the various assumptions for returns such as iid returns, non-iid returns, and time aggregations.

dependencies resulting from heteroscedasticity or serial correlation (see Kosowski *et al.*, 2007).

Essentially, our purpose in using the bootstrap method is to determine whether the performance of each mutual fund and its correlation with market returns hinges on sample variability. Therefore, to obtain the distribution of the Sharpe ratio and the multiscale correlations, we re-sample the vector of mutual funds and market returns. To be more specific, at the first stage, we re-sample 3,000 replications of the vector of mutual fund and market returns using the bootstrap approach. At the second stage, we reproduce the same results with the original data series using a wavelet multiscaling approach and, using these reproduced results, construct the confidence intervals for the Sharpe ratio and the multiscale correlations.

10.3. Data and Empirical Results

We use monthly nominal mutual fund returns (index, institutional, and active funds) for the US for the period January 1991 – December 2005. Data were collected from CRSP. To construct the returns of each fund group, their value-weighted returns are calculated using total net assets under management. More specifically, for the index fund category there are 12 sampled funds and their returns are weight averaged using their net asset values. Similarly, institutional fund and active fund group returns are calculated from 35 institutional funds and 346 active funds, respectively. For the market return proxy, we use the CRSP value-weighted market index return; for the risk free return, we use one-month Treasury bills.

Institutional funds are defined as mutual funds that target pension funds, endowments, and other high net worth entities and individuals. Institutional funds usually have lower operating costs and higher minimum initial investments than retail funds. The main objective of institutional funds is to reduce risk by investing in hundreds of different securities. The objective of Active funds is to outperform the market average by actively seeking out stocks that will provide superior total return. In contrast, Index funds are a form of passive investment. Index funds are mutual funds whose portfolio aims to match the holdings (with some degree of tolerance on tracking error) of a market index, such as the S&P500 Index. Therefore, their performance mirrors the market as a whole.

Table 10.1 presents several summary statistics for the monthly data from our three groupings of mutual funds (Institutional, Active and

Table 10.1.   Descriptive statistics.

	Market	Index	Institutional	Active
Mean	1.0200	0.9850	1.0800	0.9999
Std.dev	4.1500	4.0500	3.5100	4.0200
Skewness	−0.6870	−0.4951	−0.8406	−0.6086
Kurtosis	1.1292	0.9026	2.2494	1.1020
LB(5)	1.1987	3.7449	3.1053	0.9449
	(0.5492)	(0.1537)	(0.2117)	(0.6235)
LB(10)	6.5884	8.6028	5.6759	5.2442
	(0.4730)	(0.2824)	(0.5781)	(0.6302)
$LB^2(5)$	10.5675*	12.7537*	1.7071	7.1816*
	(0.0051)	(0.0017)	(0.4259)	(0.0276)
$LB^2(10)$	23.5555*	22.0862*	9.4895	16.9636*
	(0.0014)	(0.0025)	(0.2194)	(0.0176)
$\rho(t, t-1)$	0.0032	−0.0604	0.0499	−0.0103

Note: Institutional funds are mutual funds which target pension funds, endowments, and other high net worth entities and individuals. The objective of active funds is to outperform the market average by actively seeking out stocks that will provide superior total return. Index funds are a class of mutual funds whose portfolio matches that of a market index, such as the S&P 500 Index. "Market" is the CRSP value weighted returns. *indicates significance at the 5% level. LB(k) and $LB^2(k)$ denote the Ljung–Box test of significance of autocorrelations of k lags for returns and squared returns, respectively. $\rho(t, t-1)$ is the first order autocorrelation coefficient. Skewness and kurtosis are defined as $E[(R_t - \mu)]^3$, and $E[(R_t - \mu)]^4$, respectively, where μ is the sample mean.

Index) — their returns and market returns, indicated as MKT. As shown in Table 10.1, all sample means range from 0.985 (Index) to 1.080 (Institutional). Comparing the three fund categories, all have very similar mean returns and standard deviation of returns, with a slightly lower standard deviation for Institutional funds. Table 10.1 also reveals that across the four reported variables, first-order autocorrelation of monthly data ranges from −0.060 (Index) to 0.050 (Institutional), implying that Institutional funds are more persistent than the two mutual fund categories and the market portfolio. The Ljung-Box statistics indicate the persistence of linear dependency for each set of data; the Ljung-Box statistics for the squared data show strong evidence of non-linear dependency in all data except in that for Institutional funds. The measures for skewness and kurtosis are also reported in order to determine whether monthly data are normally distributed; these statistics indicate that all data are not normally distributed.

10.3.1. *Estimation results for the three aggregate groups*

The purpose of this chapter is to examine the performance of mutual funds across multiple time horizons. To do so, we use a multi-horizon version of the Sharpe ratio, employing a wavelet multiscaling approach. Considering the balance between the sample size and the length of the wavelet filter, we employ the Daubechies extremal phase wavelet filter of length 4 [D(4)], while we decompose our data up to scale 5. Since we use monthly data, scale 1 represents 1–2 month period dynamics. Equivalently, scales 2, 3, 4, and 5 correspond to 2–4, 4–8, 8–16, and 16–32 month period dynamics, respectively.

First, we examine the variances of the three groups of mutual fund returns against various time scales. The examination of variance at each time scale is important in our study because such variance shows how much risk investors incur at that time scale in the context of the Sharpe ratio. An important characteristic of the wavelet transform is its ability to decompose (for analytical purposes) the variance of the stochastic process. Fig. 10.1 illustrates the wavelet variance of the three series against the wavelet scales. The return variances of the three mutual fund categories decrease as the wavelet scale increases. Note that the variance-versus-wavelet scale curves show a broad peak at the lowest scale (scale 1) in all groups. This result implies that an investor with a short investment horizon must respond to every fluctuation in realized returns, while for an investor with a much longer horizon, the long-run risk is significantly lower (Kim and In, 2005a).

Table 10.2 presents the multiscale correlation with market portfolios. Investigating correlation with market returns is important for the construction of a portfolio of investments because inclusion of mutual funds with low correlations can provide better risk-return tradeoffs. Note that the mean and the upper and lower bounds at the 5% significance level are calculated using the nonparametric bootstrap method by generating 3,000 replications. As shown in Table 10.2, the average across replications is very similar to the original estimates, which means that no correction for small-sample bias is needed. Most correlations, regardless of time scale, show very high correlation with market returns, while Institutional funds at scale 5 show somewhat lower correlation with market returns, implying that inclusion of an additional mutual fund in a portfolio is not a good strategy for all investors across various investment horizons.

In Table 10.3, we report the estimated Sharpe ratio and the average Sharpe ratio, and their corresponding standard errors, calculated using the nonparametric bootstrap method. Table 10.3 also presents the Sharpe ratio

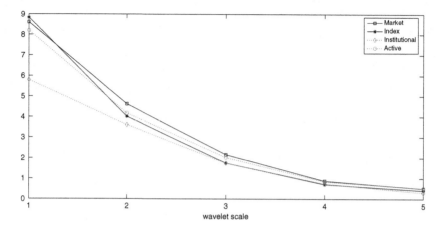

Fig. 10.1.   Estimated wavelet variances.

Note: The wavelet scales are as follows: scale 1, 1–2 month period dynamics; scale 2, 2–4 month period dynamics; scale 3, 4–8 month period dynamics; scale 4, 8–16 month period dynamics; scale 5, 16–32 month period dynamics.

Table 10.2.   Estimated correlations.

	Panel A. Estimated correlations			Panel B. Average correlations		
	Index	Institutional	Active	Index	Institutional	Active
Raw	0.969	0.956	0.988	0.945	0.918	0.963
Scale 1	0.977	0.971	0.995	0.977	0.971	0.995
Scale 2	0.969	0.965	0.994	0.969	0.965	0.994
Scale 3	0.973	0.955	0.992	0.972	0.955	0.992
Scale 4	0.975	0.922	0.989	0.975	0.920	0.989
Scale 5	0.980	0.795	0.990	0.980	0.796	0.990
	Panel C. Lower bound			Panel D. Upper bound		
	Index	Institutional	Active	Index	Institutional	Active
Raw	0.906	0.868	0.925	0.985	0.967	1.000
Scale 1	0.964	0.959	0.993	0.990	0.982	0.997
Scale 2	0.951	0.951	0.991	0.987	0.978	0.996
Scale 3	0.958	0.937	0.988	0.987	0.972	0.995
Scale 4	0.964	0.887	0.985	0.985	0.954	0.993
Scale 5	0.972	0.739	0.986	0.987	0.853	0.993

Note: To calculate correlation at scale λ_j, we decompose each time series up to level 5 using the Daubechies extremal phase wavelet filter of length 4 [D(4)]. Scales 1, 2, 3, 4, and 5 represent 1–2, 2–4, 4–8, 8–16, and 16–32 month period dynamics, respectively. The upper and lower bounds indicate the 95% confidence interval. The average correlation, lower, and upper bounds are calculated using the bootstrap method, by generating 3,000 replications.

Table 10.3.   Estimated Sharpe ratios for aggregate funds.

	Market	Index	Institutional	Active
Raw	0.170	0.165	0.217	0.170
	(0.243)	(0.228)	(0.322)	(0.242)
	[0.109]	[0.106]	[0.112]	[0.109]
Scale 1	0.220	0.208	0.293	0.216
	(0.221)	(0.208)	(0.294)	(0.216)
	[0.076]	[0.071]	[0.083]	[0.075]
Scale 2	0.298	0.306	0.370	0.300
	(0.300)	(0.309)	(0.372)	(0.303)
	[0.075]	[0.076]	[0.074]	[0.077]
Scale 3	0.425	0.462	0.525	0.416
	(0.428)	(0.464)	(0.529)	(0.419)
	[0.084]	[0.092]	[0.081]	[0.084]
Scale 4	0.661	0.742	0.774	0.618
	(0.667)	(0.749)	(0.782)	(0.624)
	[0.123]	[0.142]	[0.104]	[0.117]
Scale 5	0.422	0.439	0.707	0.394
	(0.421)	(0.439)	(0.709)	(0.394)
	[0.165]	[0.201]	[0.131]	[0.157]

Note: To calculate the Sharpe ratio at scale λ_j, we decompose each time series up to level 5, using the Daubechies extremal phase wavelet filter of length 4 [D(4)]. Scales 1, 2, 3, 4, and 5 represent 1–2, 2–4, 4–8, 8–16, and 16–32 month period dynamics, respectively. The mean values of the Sharpe ratio for the returns of each portfolio are reported in parentheses, while the standard deviations are reported in brackets. The mean values and standard errors are calculated using the bootstrap method, by generating 3,000 replications.

of the market portfolio for purposes of comparison with other mutual funds. Three features of this table are noteworthy. First, the Sharpe ratio shows its highest value at scale 4 in all portfolios, indicating that investors receive the highest compensation for a unit of risk over the 8–16 month period, i.e., approximately one year. Second, while the Sharpe ratio for Index fund raw data is lower than that for the market portfolio, the Sharpe ratio of Index funds is consistently higher than the market portfolio, indicating that the performance of Index funds is better than the market portfolio. Finally, Institutional (Active) funds have the highest (lowest) Sharpe ratio at all time scales, indicating that Institutional (Active) funds are the best (worst) performer among mutual fund categories. This final result indicates that the winner (loser) at short time scales consistently outperforms (underperforms).

In sum, our results show that Institutional funds tend to outperform Index and Active funds at all time scales, suggesting that the persistence of such out-performance exists at the time scales.

10.3.2. *Estimation results for individual mutual funds*

To examine the performance of individual funds, we select the bottom five and top five funds for each group. These funds are selected based on their average net asset values during 2005. More specifically, *bottom 1* (*top 1*) in each fund group indicates that fund with the lowest (highest) average net asset value during 2005. Estimated Sharpe ratios, average Sharpe ratios, and corresponding standard errors for the Index fund group are reported in Table 10.4. Note that as in Table 10.3, the average Sharpe ratios (reported in parentheses) and the standard errors (reported in brackets) are calculated using the nonparametric bootstrap method by generating 3,000 replications.

Overall, the Sharpe ratio at scale 4 shows the highest value in all portfolios, indicating that investors receive the highest compensation for a unit of risk over the 8–16 month period, as shown in Table 10.3. Comparison of the Sharpe ratios at each time scale shows that the size of a fund does not play an important role in its performance. For example, *top 5* (*bottom 1*) always has the highest (lowest) Sharpe ratio at all time scales, indicating that the winner (loser) at short time scales consistently outperforms (underperforms). However, the estimated Sharpe ratios are statistically significantly different from each other, because the 95% confidence interval of *bottom 1*'s Sharpe ratio includes all estimated Sharpe ratios.

Next, we examine individual funds within the Institutional fund group, as illustrated in Table 10.5. Overall, the estimated Sharpe ratios of those funds with lower net asset values are lower than funds with higher net asset values at all time scales. That is, the performance of funds with higher net asset values is consistently better than for funds with lower net asset values, implying that the size of funds matters. Note that from the bootstrapped results, the estimated Sharpe ratios between *bottom* funds and *top* funds are statistically significantly different using the 95% confidence interval. In addition, it is found that the estimated Sharpe ratios for funds with lower net asset values increase up to scale 4, while those for funds with higher net asset values increase with the time scale. This indicates that institutional funds with lower net asset values provide the highest compensation for

Table 10.4. Estimated Sharpe ratios for selected individual index funds.

	Bottom 1	Bottom 2	Bottom 3	Bottom 4	Bottom 5	Top 5	Top 4	Top 3	Top 2	Top 1
Raw	0.150 (0.207) [0.105]	0.157 (0.217) [0.106]	0.164 (0.226) [0.106]	0.160 (0.221) [0.106]	0.163 (0.225) [0.106]	0.172 (0.242) [0.108]	0.164 (0.226) [0.106]	0.164 (0.226) [0.106]	0.168 (0.232) [0.106]	0.166 (0.229) [0.106]
Scale 1	0.187 (0.188) [0.070]	0.197 (0.198) [0.071]	0.206 (0.207) [0.071]	0.200 (0.201) [0.071]	0.205 (0.206) [0.071]	0.222 (0.223) [0.074]	0.205 (0.206) [0.071]	0.206 (0.207) [0.071]	0.211 (0.212) [0.071]	0.208 (0.209) [0.071]
Scale 2	0.277 (0.280) [0.076]	0.291 (0.294) [0.077]	0.303 (0.306) [0.076]	0.296 (0.299) [0.076]	0.302 (0.305) [0.076]	0.318 (0.321) [0.078]	0.302 (0.305) [0.076]	0.303 (0.306) [0.076]	0.312 (0.315) [0.076]	0.307 (0.310) [0.076]
Scale 3	0.416 (0.418) [0.091]	0.433 (0.436) [0.091]	0.456 (0.459) [0.092]	0.446 (0.448) [0.092]	0.456 (0.458) [0.093]	0.480 (0.482) [0.094]	0.455 (0.458) [0.092]	0.457 (0.460) [0.092]	0.470 (0.473) [0.092]	0.463 (0.466) [0.092]
Scale 4	0.672 (0.678) [0.140]	0.697 (0.703) [0.140]	0.736 (0.743) [0.142]	0.719 (0.725) [0.142]	0.729 (0.736) [0.142]	0.769 (0.775) [0.143]	0.733 (0.740) [0.142]	0.735 (0.741) [0.142]	0.756 (0.762) [0.142]	0.744 (0.751) [0.142]
Scale 5	0.357 (0.356) [0.200]	0.384 (0.384) [0.202]	0.428 (0.428) [0.201]	0.396 (0.396) [0.201]	0.418 (0.418) [0.200]	0.484 (0.483) [0.211]	0.430 (0.430) [0.201]	0.428 (0.427) [0.201]	0.457 (0.457) [0.201]	0.443 (0.443) [0.201]

Note: To calculate the Sharpe ratio at scale λ_j, we decompose each time series up to level 5, using the Daubechies extremal phase wavelet filter of length 4 [D(4)]. Scales 1, 2, 3, 4, and 5 represent 1–2, 2–4, 4–8, 8–16, and 16–32 month period dynamics, respectively. The mean values of the Sharpe ratio for the returns of each portfolio are reported in parentheses, while the standard errors are reported in brackets. The mean values and standard deviations are calculated using the bootstrap method, by generating 3,000 replications.

Table 10.5. Estimated Sharpe ratios for selected individual institutional funds.

	Bottom 1	Bottom 2	Bottom 3	Bottom 4	Bottom 5	Top 5	Top 4	Top 3	Top 2	Top 1
Raw	0.091 (0.132) [0.111]	0.063 (0.091) [0.110]	0.211 (0.304) [0.111]	0.132 (0.194) [0.113]	0.191 (0.278) [0.110]	0.194 (0.270) [0.105]	0.288 (0.457) [0.121]	0.283 (0.449) [0.121]	0.159 (0.228) [0.109]	0.280 (0.436) [0.119]
Scale 1	0.128 (0.127) [0.078]	0.084 (0.083) [0.077]	0.280 (0.282) [0.078]	0.171 (0.171) [0.079]	0.254 (0.256) [0.081]	0.244 (0.245) [0.072]	0.426 (0.428) [0.097]	0.417 (0.419) [0.097]	0.193 (0.195) [0.075]	0.406 (0.407) [0.094]
Scale 2	0.159 (0.160) [0.071]	0.118 (0.120) [0.090]	0.375 (0.378) [0.075]	0.213 (0.215) [0.074]	0.338 (0.341) [0.075]	0.368 (0.371) [0.077]	0.496 (0.501) [0.083]	0.475 (0.479) [0.080]	0.280 (0.283) [0.073]	0.473 (0.477) [0.080]
Scale 3	0.220 (0.222) [0.081]	0.154 (0.154) [0.103]	0.524 (0.528) [0.080]	0.300 (0.303) [0.080]	0.477 (0.480) [0.082]	0.551 (0.554) [0.091]	0.687 (0.694) [0.089]	0.666 (0.673) [0.088]	0.423 (0.425) [0.086]	0.678 (0.685) [0.089]
Scale 4	0.255 (0.260) [0.090]	0.178 (0.181) [0.131]	0.916 (0.925) [0.125]	0.463 (0.470) [0.106]	0.622 (0.626) [0.092]	0.815 (0.821) [0.118]	0.895 (0.903) [0.093]	0.850 (0.857) [0.091]	0.642 (0.646) [0.098]	0.847 (0.852) [0.084]
Scale 5	0.079 (0.079) [0.086]	−0.138 (−0.141) [0.149]	0.895 (0.897) [0.181]	0.279 (0.280) [0.087]	0.567 (0.569) [0.113]	0.721 (0.723) [0.193]	1.049 (1.058) [0.110]	0.949 (0.957) [0.097]	0.647 (0.651) [0.157]	0.944 (0.950) [0.097]

Note: To calculate the Sharpe ratio at scale λ_j, we decompose each time series up to level 5, using the Daubechies extremal phase wavelet filter of length 4 [D(4)]. Scales 1, 2, 3, 4, and 5 represent 1–2, 2–4, 4–8, 8–16, and 16–32 month period dynamics, respectively. The mean values of the Sharpe ratio for the returns of each portfolio are reported in parentheses, while the standard errors are reported in brackets. The mean values and standard deviations are calculated using the bootstrap method, by generating 3,000 replications.

Table 10.6. Estimated Sharpe ratios for selected individual active funds.

	Bottom 1	Bottom 2	Bottom 3	Bottom 4	Bottom 5	Top 5	Top 4	Top 3	Top 2	Top 1
Raw	−0.187 (−0.290) [0.120]	−0.032 (−0.047) [0.109]	0.110 (0.176) [0.118]	0.057 (0.079) [0.106]	0.127 (0.188) [0.114]	0.252 (0.361) [0.108]	0.260 (0.380) [0.115]	0.201 (0.279) [0.106]	0.206 (0.283) [0.105]	0.197 (0.284) [0.112]
Scale 1	−0.288 (−0.293) [0.091]	−0.054 (−0.056) [0.076]	0.135 (0.135) [0.090]	0.076 (0.076) [0.072]	0.169 (0.169) [0.082]	0.335 (0.337) [0.079]	0.339 (0.340) [0.080]	0.260 (0.262) [0.073]	0.257 (0.259) [0.071]	0.255 (0.256) [0.076]
Scale 2	−0.360 (−0.363) [0.093]	−0.073 (−0.073) [0.080]	0.164 (0.165) [0.073]	0.109 (0.110) [0.072]	0.220 (0.222) [0.083]	0.455 (0.459) [0.074]	0.476 (0.480) [0.090]	0.372 (0.375) [0.071]	0.391 (0.395) [0.075]	0.359 (0.361) [0.080]
Scale 3	−0.591 (−0.595) [0.107]	−0.151 (−0.152) [0.091]	0.211 (0.211) [0.074]	0.144 (0.144) [0.081]	0.270 (0.272) [0.071]	0.676 (0.680) [0.078]	0.561 (0.564) [0.086]	0.609 (0.612) [0.093]	0.570 (0.574) [0.085]	0.494 (0.496) [0.083]
Scale 4	−0.801 (−0.803) [0.144]	−0.284 (−0.284) [0.122]	0.288 (0.290) [0.073]	0.272 (0.277) [0.120]	0.406 (0.408) [0.071]	0.965 (0.970) [0.083]	0.777 (0.781) [0.111]	0.805 (0.811) [0.105]	0.946 (0.955) [0.128]	0.762 (0.768) [0.114]
Scale 5	−0.651 (−0.655) [0.150]	−0.297 (−0.300) [0.127]	0.530 (0.533) [0.059]	0.061 (0.061) [0.127]	0.582 (0.584) [0.105]	1.558 (1.568) [0.131]	0.673 (0.675) [0.156]	0.883 (0.888) [0.158]	1.089 (1.094) [0.212]	0.693 (0.696) [0.129]

Note: To calculate the Sharpe ratio at scale λ_j, we decompose each time series up to level 5, using the Daubechies extremal phase wavelet filter of length 4 [D(4)]. Scales 1, 2, 3, 4, and 5 represent 1–2, 2–4, 4–8, 8–16, and 16–32 month period dynamics, respectively. The mean values of the Sharpe ratio for the returns of each portfolio are reported in parentheses, while the standard errors are reported in brackets. The mean values and standard deviations are calculated using the bootstrap method, by generating 3,000 replications.

bearing a unit of risk over a one-year period, while institutional funds with higher net asset values provide compensation that increases with the time scale. Therefore, pension funds, whose investment horizon is longer than one-year, could benefit from investing in institutional funds with higher net asset values. Furthermore, overall fund ranking, calculated using estimated Sharpe ratios, does not change across time scales, indicating that in the absence of calculating the Sharpe ratio for their specific investment horizon, investors choose their investments using the Sharpe ratio at the shortest investment horizon.

Estimated Sharpe ratios for Active funds are reported in Table 10.6. Overall results are similar to those of Institutional funds. As indicated above, active funds attempt to outperform the market average by actively seeking out stocks that will provide superior total returns. Comparing the estimated Sharpe ratios of market returns, shown in Table 10.3, individual active funds with higher (lower) net asset values consistently perform better (worse) than the market return. This result suggests that investors seeking superior returns could benefit from investing in active funds with higher net asset values.

10.4. Summary and Concluding Remarks

Despite its importance in modern financial analysis, the evaluation of mutual fund performance has not been accompanied by examination of the impact of investment horizon, an important factor for investments. This chapter uses the Sharpe ratio across various time scales to evaluate the performance of three groups of US mutual funds (Institutional, Active and Index). The wavelet multiscaling approach is capable of decomposing time series across various time scales, a useful and advantageous feature that facilitates a modeling of the behavior of data within multiple time horizons.

In terms of performance measures for the three mutual fund groups, our empirical results indicate that Institutional funds are clearly dominant across all time scales. Since risk and value (performance) are timescale-dependent concepts, any form of measurement, such as the frequently used Sharpe ratio or Jensen's alpha, must account for any investment horizon effect.

For any particular Index fund, it is found that the size of the fund has no impact on performance; for Institutional and Active funds, funds with higher net asset values consistently perform better than those with lower net asset values, implying that investors seeking superior returns could benefit from investing in funds with higher net asset values.

References

Abry, P. A., D. Veitch and P. Flandrin (1998). Long range dependence: Revisiting aggregation with wavelets. *Journal of Time Series Analysis*, 19, 253–266.

Agarwal, S., C. Liu and S. G. Rhee (2007). Where does price discovery occur for stocks traded in multiple markets? Evidence from Hong Kong and London. *Journal of International Money and Finance*, 26, 46–63.

Al Awad, M. and B. K. Goodwin (1998). Dynamic linkages among real interest rates in international capital markets. *Journal of International Money and Finance*, 17, 881–907.

Albrecht, T. (1998). The mean-variance framework and long horizons. *Financial Analysts Journal*, 54, 44–49.

Asimakopoulos, I., J. Goddard and C. Siriopoulos (2000). Interdependence between the US and major European equity markets: Evidence from spectral analysis. *Applied Financial Economics*, 10, 41–47.

Bacchetta, P. and E. van Wincoop (2006). Can information heterogeneity explain the exchange rate determination puzzle? *American Economic Review*, 96, 552–576.

Bae, K. H., B. Cha and Y. L. Cheung (1999). The transmission of pricing information of dually-listed stocks. *Journal of Business Finance and Accounting*, 26, 709–723.

Ball, R. (1989). What do we know about stock market efficiency? In Guimaraes, R. M., C. B. Kingsman and S. Taylor (eds.), *A Reappraisal of the Efficiency of Financial Markets*. Heidelberg: Springer-Verlag.

Barkoulas, J. T., W. C. Labys and J. I. Onochie (1997). Fractional dynamics in international commodity prices. *Journal of Futures Markets*, 2, 737–745.

Barnes, M., J. H. Boyd and B. D. Smith (1999). Inflation and asset returns. *European Economic Review*, 43, 737–754.

Barrow, S. (1994). Fundamental currency forecasting: An agnostic view. Working Paper, *Financial Economics*, 3, 8–12.

Baxter, M. and R. G. King (1999). Measuring business cycles: Approximate band-pass filters for economic time series. *The Review of Economics and Statistics*, 81, 573–593.

Best, R., C. W. Hodges and J. Yoder (2007). The Sharpe ratio and long-run investment decisions. *Journal of Investing*, 16, 70–76.

Biddle, G. and S. Saudagaran (1989). The effects of financial disclosure levels on firms choices among alternative foreign stock exchange listings. *Journal of International Financial Management and Accounting*, 1, 55–87.

Booth, G., F. R. Kaen and P. E. Koveos (1982a). Persistent dependence in gold prices. *Journal of Financial Research*, 5(1), 85–93.

Booth, G., F. R. Kaen and P. E. Koveos (1982b). R/S analysis of foreign exchange rates under two international monetary regimes. *Journal of Monetary Economics*, 10, 407–415.

Boudoukh, J. and M. Richardson (1993). Stock returns and inflation: A long horizon perspective. *American Economic Review*, 83, 1346–1355.

Brooks, C., O. T. Henry and G. Persand (2002). The effect of asymmetries on optimal hedge ratios. *Journal of Business*, 75, 333–352.

Brown, K. C., W. V. Harlow and D. J. Smith (1994). An empirical analysis of interest rate swap spreads. *Journal of Fixed Income*, 3, 61–68.

Brown, R., F. In and V. Fang (2002). Modeling the determinants of swap spreads. *Journal of Fixed Income*, 12, 29–40.

Brown, S. J., W. N. Goetzmann, R. G. Ibbotson and S. A. Ross (1992). Survivorship bias in performance studies. *Review of Financial Studies*, 5, 553–580.

Burrus, C. S., R. A. Gopinath and H. Guo (1998). *Introduction to Wavelets and Wavelet Transforms*. New Jersey: Prentice-Hall.

Cao, C., E. Ghysels and F. Hatheway (2000). Price discovery without trading: Evidence from the Nasdaq preopening. *Journal of Finance*, 55, 1339–1365.

Capobiance, E. (2003). Empirical volatility analysis: Feature detection and signal extraction with function dictionaries. *Physica A*, 319, 495–518.

Cech, C. (2006). Copula-based top-down approaches in financial risk aggregation. The University of Applied Sciences of bfi Vienna, Working Paper Series 32.

Chang, R. P., H. Oppenheimer and S. G. Rhee (1997). The offshore migration of trading of Hong Kong-listed stocks: Causes and effects. Hong Kong: Hong Kong securities and futures commission.

Chen, N. F., R. Roll and S. A. Ross (1986). Economic forces and the stock market. *Journal of Business*, 59, 383–403.

Chen, S. S., C. F. Lee and K. Shrestha (2001). On a mean-generalized semivariance approach to determining the hedge ratio. *Journal of Futures Markets*, 21, 581–598.

Cheung, Y. L. and C. K. Shum (1995). International stock exchange listing and the reduction of political risk. *Managerial and Decision Economics*, 16, 537–546.

Cheung, Y. W. and F. X. Diebold (1994). On maximum likelihood estimation of the differencing parameter of fractionally integrated noise with unknown mean. *Journal of Econometrics*, 62, 301–316.

Cheung, Y. W. and K. S. Lai (1993). Do gold market returns have long memory? *The Financial Review*, 28, 181–202.

Chew, C. (2001). The money and income relationship of European countries by time scale decomposition using wavelets. Unpublished Paper, New York University.

Chopra, V. K. and W. T. Ziemba (1993). The effect of error in means, variances and covariances on optimal portfolio choice. *Journal of Portfolio Management*, 19, 6–11.

Choudhry, T. (2001). Inflation and rates of return on stocks: Evidence from high inflation countries. *Journal of International Financial Markets, Institutions and Money*, 11, 75–96.

Choudhry, T. (2003). Short-run deviations and optimal hedge ratio: Evidence from stock futures. *Journal of Multinational Financial Management*, 13, 171–192.

Chui, C. K. (1992). *An Introduction to Wavelets*. San Diago: Academic Press.

Collin-Dufresne, P. and B. Solnik (2001). On the term structure of default premia in the swap and LIBOR markets. *Journal of Finance*, 61, 1095–1115.

Conner, J. and R. Rossiter (2005). Wavelet transform and commodity prices. *Studies in Nonlinear Dynamics and Econometrics*, 9, Article 6.

Cooper, M. J., R. C. Gutierrez Jr. and A. Hameed (2004). Market states and momentum. *Journal of Finance*, 59, 1345–1365.

Crato, N. and B. K. Ray (2000). Memory in returns and volatilities of futures' contracts. *Journal of Futures Markets*, 20, 525–543.

Cvitanić, J., A. Lazrak and T. Wang (2008). Implications of the Sharpe ratio as a performance measure in multi-period settings. *Journal of Economic Dynamics and Control*, 32, 1622–1649.

Daniel, K. and S. Titman (2006). Market reactions to tangible and intangible information. *Journal of Finance*, 61, 1605–1643.

Daniel, K., D. Hirsleifer and A. Subrahmanyam (1998). Investor psychology and investor security market under and overreaction. *Journal of Finance*, 53, 1839–1886.

Daubechies, I. (1992). *Ten Lectures on Wavelets*. Philadelphia: SIAM.

Davidson, R., W. C. Labys and J. B. Lesourd (1998). Wavelet analysis of commodity price behavior. *Computational Economics*, 11, 103–128.

Doidge, C., A. Karolyi and R. Stulz (2004). Why are foreign firms listed in the US worth more? *Journal of Financial Economics*, 71, 205–238.

Duffie, D. (1989). *Futures Markets*. Englewood Cliffs, NJ: Prentice-Hall.

Duffie, D. and K. J. Singleton (1997). An econometric model of the term structure of interest-rate swap yields. *Journal of Finance*, 52, 1287–1321.

Duffie, D. and M. Huang (1996). Swap rates and credit quality. *Journal of Finance*, 51, 921–949.

Ederington, L. (1979). The hedging performance of the new futures markets. *Journal of Finance*, 34, 15–7170.

Efron, B. (1979). Bootstrap methods: Another look at the Jackknife. *Annals of Statistics*, 7, 1–26.

Eling, M. (2008). Does the measure matter in the mutual fund industry? *Financial Analysts Journal*, 64, 1–13.

Eling, M. and F. Schuhmacher (2007). Does the choice of performance measure influence the evaluation of hedge funds? *Journal of Banking and Finance*, 31, 2632–2647.

Ely, D. P. and K. J. Robinson (1997). Are stocks a hedge against inflation? International evidence using a long-run approach. *Journal of International Money and Finance*, 16, 141–167.

Eom, Y. H., M. G. Subrahmanyam and J. Uno (2002). The international linkages of interest rate swap spreads: The yen–dollar markets. *Journal of Fixed Income*, 12, 6–28.

Eun, S. C. and S. Sabherwal (2003). Cross-boarder listings and price discovery: Evidence from US listed Canadian stocks. *Journal of Finance*, 58, 549–575.

Fama, E. F. and G. W. Schwert (1977). Stock market returns and inflation. *Journal of Financial Economics*, 5, 115–146.

Fama, E. and K. French (1992). The cross-section of expected stock returns. *Journal of Finance*, 47, 427–465.

Fama, E. and K. French (1993). Common risk factors in the returns on stocks and bonds. *Journal of Financial Economics*, 33, 3–56.

Fama, E. and K. French (1995). Size and book-to-market factors in earnings and returns. *Journal of Finance*, 50, 131–156.

Fama, E. and K. French (1996). Multifactor explanations of asset pricing anomalies. *Journal of Finance*, 51, 55–84.

Fama, E. and K. French (1997). Industry costs of equity. *Journal of Financial Economics*, 43, 153–193.

Fernandez, V. P. (2005). The international CAPM and a wavelet-based decomposition of Value at Risk. *Studies in Nonlinear Dynamics and Econometrics*, 9, Article 4.

Fischer, K. P. and A. P. Palasvirta (1990). High road to a global marketplace: The international transmission of stock market fluctuations. *The Financial Review*, 25, 371–393.

Fleming, B. J. W., D. Yu and R. G. Harrison (2000). Analysis of effect of detrending of time-scale structure of financial data using discrete wavelet transform. *International Journal of Theoretical and Applied Finance*, 3, 357–379.

Gabor, D. (1946). Theory of communication. *Journal of the IEE*, 93, 429–457.

Gençay, R., F. Selçuk and B. Whitcher (2001). Scaling properties of foreign exchange volatility. *Physica A*, 289, 249–266.

Gençay, R., F. Selçuk and B. Whitcher (2002). Robustness of systematic risk across time scales. Working Paper.

Gençay, R., F. Selçuk and B. Whitcher (2002). *An Introduction to Wavelets and Other Filtering Methods in Finance and Economics*. London: Academic Press.

Gençay, R., F. Selçuk and B. Whitcher (2003). Systematic risk and time scales. *Quantitative Finance*, 3, 108–116.

Gençay, R., F. Selçuk and B. Whitcher (2005). Multiscale systematic risk. *Journal of International Money and Finance*, 24, 55–70.

Gençay, R., N. Gradojevic, F. Selçuk and B. Whitcher (2010). Asymmetry of information flow between volatilities across time scales. *Quantitative Finance*, 10, 895–915.

Geppert, J. M. (1995). A statistical model for the relationship between futures contract hedging effectiveness and investment horizon length. *Journal of Futures Markets*, 15, 507–536.

Geweke, J. and S. Porter-Hudak (1983). The estimation and application of long memory time series models. *Journal of Time Series Analysis*, 4(4), 221–238.

Goffe, W. L. (1994). Wavelets in macroeconomics: An introduction. In Belsley, D. A. (ed.), *Computational Techniques for Econometrics and Economic Analysis*. Boston, MA: Kluwer Academic Publishers, pp. 137–149.

Graps, A. (1995). An introduction to wavelets. *IEEE Computational Science and Engineering*, 2, 50–61.

Grossman, A. and J. Morlet (1984). Decomposition of Hardy functions into square integrable wavelets of constant shape. *SIAM Journal of Mathematical Analysis*, 15, 723–736.

Hall, G. J. and S. Krieger (2000). The tax smoothing implications of the federal debt paydown. Brookings Papers on Economic Activity 2000:2.

Hallett, A. H. and C. R. Richter (2001). A comparative dynamic analysis of British and German monetary policy in the 1990s: A spectral analysis approach. Working Paper.

Hallett, A. H. and C. R. Richter (2002). Are capital markets efficient? Evidence from the term structure of interest rates in Europe. *The Economic and Social Review*, 33, 333–356.

Handa, P., S. P. Kothari and C. Wasley (1989). The relation between the return interval and betas: Implications for the size effect. *Journal of Financial Economics*, 23, 79–100.

Hansson, B. and M. Persson (2000). Time diversifications and estimation risk. *Financial Analysts Journal*, 56, 55–62.

Harrison, P. and H. H. Zhang (1999). An investigation of the risk and return relation at long horizons. *Review of Economics and Statistics*, 81, 399–408.

Hill, J. and T. Schneeweis (1982). The hedging effectiveness of foreign currency futures. *Journal of Financial Research*, 5, 95–104.

Hodges, C. W., W. R. Taylor and J. A. Yoder (1997). Stocks, bonds, the Sharpe ratio, and the investment horizon. *Financial Analysts Journal*, 53, 74–80.

Holton, G. A. (1992). Time: The second dimension of risk. *Financial Analysts Journal*, 48, 38–45.

Hong, H. and J. Stein (1999). A unified theory of underreaction, momentum trading, and overreaction in asset markets. *Journal of Finance*, 54, 2143–2184.

Howard, C. T. and L. J. D'Antonio (1984). A risk-return measure of hedging effectiveness. *Journal of Financial and Quantitative Analysis*, 19, 101–112.

Howard, C. T. and L. J. D'Antonio (1991). Multiperiod hedging using futures: A risk minimization approach in the presence of autocorrelation. *Journal of Futures Markets*, 11, 697–710.

Howrey, E. P. (1968). A spectrum analysis of the long-swing hypothesis. *International Economic Review*, 9, 228–252.

Hsin, C., W. Kuo and C. F. Lee (1994). A new measure to compare the hedging effectiveness of foreign currency futures versus options. *Journal of Futures Markets*, 14, 685–707.

Hudgins, L., C. A. Friehe and M. E. Mayer (1993). Wavelet transforms and atmospheric turbulence. *Physical Review Letters*, 71, 3279–3282.

Hurst, H. E. (1951). Long-term storage capacity of reservoirs. *Transactions of the American Society of Civil Engineers*, 116, 770–779.

Hurvich, C. and K. Beltrao (1993). Asymptotics for the low-frequency ordinates of the periodogram of a long-memory time series. *Journal of Time Series Analysis*, 14, 455–472.

In, F. and S. Kim (2006). The hedge ratio and the empirical relationship between the stock and futures markets: A new approach using wavelet analysis. *Journal of Business*, 79, 799–820.

In, F., R. Brown and V. Fang (2003a). Links among interest rate swap markets: US, UK, and Japan. *Journal of Fixed Income*, 13, 84–95.

In, F., R. Brown and V. Fang (2003b). Modeling volatility and changes in the swap spread. *International Review of Financial Analysis*, 12, 545–561.

In, F., R. Brown and V. Fang (2004). Australian and US interest rate swap markets: Comparison and linkages. *Accounting and Finance*, 44, 45–56.

Investment Company Institute (2010). *Investment Company Fact Book*, 50th edition.

Jamdee, S. and C. A. Los (2006). Dynamic risk profile of the US term structure by wavelet MRA. *International Research Journal of Finance and Economics*, 5, 20–47.

Jensen, M. J. (1968). The performance of mutual funds in the period 1945–1964. *Journal of Finance*, 23, 389–416.

Jensen, M. J. (1999a). An approximate wavelet MLE of short and long memory parameters. *Studies in Nonlinear Dynamics and Economics*, 3, 239–253.

Jensen, M. J. (1999b). Using wavelets to obtain a consistent ordinary least square estimator of the long-memory parameter. *Journal of Forecasting*, 18, 17–32.

Jensen, M. J. (2000). An alternative maximum likelihood estimator of long-memory processes using compactly supported wavelets. *Journal of Economic Dynamics and Control*, 24, 361–387.

Johnson, L. L. (1960). The theory of hedging and speculation in commodity futures. *Review of Economic Studies*, 27, 139–151.

Kan, R. and D. R. Smith (2008). The distribution of the sample minimum-variance frontier. *Management Science*, 54, 1364–1380.

Karolyi, G. A. (1998). Why do companies list shares abroad? A survey of the evidence and its managerial implications. *Financial Markets, Institutions and Instruments*, 7, 1–60.

Kim, M., A. C. Szakmary and I. Mathur (2000). Price transmission dynamics between ADRs and their underlying foreign securities. *Journal of Banking and Finance*, 24, 1359–1382.

Kim, S. and F. In (2003). The relationship between financial variables and real economic activity: Evidence from spectral and wavelet analyses. *Studies in Nonlinear Dynamics and Econometrics*, 7(4), 1–16.

Kim, S. and F. In (2005a). Multihorizon Sharpe ratio. *Journal of Portfolio Management*, 31, 105–111.

Kim, S. and F. In (2005b). The relationship between stock returns and inflation: New evidence from wavelet analysis. *Journal of Empirical Finance*, 12, 435–444.

Kim, S. and F. In (2010). Portfolio allocation and the investment horizon: A multiscaling approach. *Quantitative Finance*, 10, 443–453.

Kirchgassner, G. and J. Wolters (1987). US–European interest rate linkage: A time series analysis for West German, Switzerland, and the United States. *Review of Economics and Statistics*, 69, 675–684.

Knif, J., S. Pynnonen and M. Luoma (1995). An analysis of lead-lag structures using a frequency domain approach: Empirical evidence from the Finnish and Swedish stock markets. *European Journal of Operational Research*, 81, 259–270.

Kosowski, R., A. Timmermann, R. Wermers and H. White (2006). Can mutual fund "stars" really pick stocks? New evidence from a bootstrap analysis. *Journal of Finance*, 61, 2551–2596.

Kosowski, R., N. Y. Naik and T. Melvyn (2007). Do hedge funds deliver alpha? A Baysian and bootstrap analysis. *Journal of Financial Economics*, 84, 229–264.

Kothari, S. and J. Shanken (1998). On defence of beta. In Stern, J. and D. Chew (eds.), *The Revolution in Corporate Finance*, 3rd Edn. New York: Blackwell, pp. 52–57.

Kothari, S. P. and J. B. Warner (2001). Evaluating mutual fund performance. *Journal of Finance*, 56, 1985–2010.

Kroner, K. F. and J. Sultan (1993). Time-varying distribution and dynamic hedging with foreign currency futures. *Journal of Financial and Quantitative Analysis*, 28, 535–551.

Kyaw, N., C. A. Los and S. Zong (2006). Persistence characteristics of Latin American financial markets. *Journal of Multinational Financial Management*, 16, 269–290.

Lakonishok, J., A. Schleifer and R. Vishny (1994). Contrarian investment, extrapolation, and risk. *Journal of Finance*, 49, 1541–1578.

Lang, L. H. P., R. H. Litzenberger and A. L. Liu (1998). Determinants of interest rate swap spreads. *Journal of Banking and Finance*, 22, 1507–1532.

Lau, S. T. and J. D. Diltz (1994). Stock returns and the transfer of information between the New York and Tokyo stock exchanges. *Journal of International Money and Finance*, 13, 211–222.

Lee, C. F., C. Wu and K. C. J. Wei (1990). The heterogeneous investment horizon and the capital asset pricing model: Theory and implications. *Journal of Financial Quantitative Analysis*, 25, 361–376.

Lee, G. G. J. (1999). Contemporary and long-run correlations: A covariance component model and studies on the S&P500 cash and futures markets. *Journal of Futures Markets*, 19, 877–894.

Lee, J. and Y. Hong (2001). Testing for serial correlation of unknown form using wavelet methods. *Econometric Theory*, 17, 386–423.

Lekkos, I. and C. Milas (2001). Identifying the factors that affect interest-rate swap spreads: Some evidence from the United States and the United Kingdom. *Journal of Futures Markets*, 21, 737–768.

Lettau, M. and S. Ludvigson (2001). Resurrecting the (C)CAPM: A cross-sectional test when risk-premia are time varying. *Journal of Political Economy*, 109, 1238–1287.

Lettau, M. and J. A. Wachter (2007). Why is long-horizon equity less risky? A duration-based explanation of the value premium. *Journal of Finance*, 62, 55–92.

Levhari, D. and H. Levy (1977). The capital asset pricing model and the investment horizon. *Review of Economics and Statistics*, 59, 92–104.

Levy, H. (1972). Portfolio performance and the investment horizon. *Management Science*, 18, 645–653.

Lewellen, J. (1999). The time-series relations among expected return, risk, and book-to-market. *Journal of Financial Economics*, 54, 5–43.

Li, H. and C. Mao (2003). Corporate use of interest rate swaps: Theory and evidence. *Journal of Banking and Finance*, 27, 1511–1538.

Lien, D. and B. K. Wilson (2001). Multiperiod hedging in the presence of stochastic volatility. *International Review of Financial Analysis*, 10, 395–406.

Lien, D. and X. Luo (1993). Estimating multiperiod hedge ratios in cointegrated markets. *Journal of Futures Markets*, 13, 909–920.

Lien, D. and X. Luo (1994). Multiperiod hedging in the presence of conditional heteroskedasticity. *Journal of Futures Markets*, 14, 927–955.

Liew, J. and M. Vassalou (2000). Can book-to-market, size and momentum be risk-factors that predict economic growth? *Journal of Financial Economics*, 57, 221–245.

Lin, S.-J. and M. Stevenson (2001). Wavelet analysis of the cost-of-carry model. *Studies in Nonlinear Dynamics and Econometrics*, 5, Article 7.

Lin, S. M., K. Craft and V. Chow (1996). Spectral analysis in three dimensions: The examination of economic interdependence between New York, London, Tokyo and the Pacific Basin equity market indices. *Journal of Applied Business Research*, 12, 72–84.

Lindsay, R. W., D. B. Percival and D. A. Rothrock (1996). The discrete wavelet transform and the scale analysis of the surface properties of sea ice. *IEEE Transactions on Geoscience and Remote Sensing*, 34, 771–787.

Lo, A. (1991). Long term memory in stock market prices. *Econometrica*, 59, 1279–1313.

Lo, A. W. (2002). The statistics of Sharpe ratios. *Financial Analysts Journal*, 58, 36–52.

Lobato, I. N. and P. M. Robinson (1998). A nonparametric test for I(0). *Review of Economic Studies*, 65, 475–495.

Los, C. A. (2003). *Financial Market Risk: Measurement and Analysis*. New York: Routledge.

Low, A., J. Muthuswamy, S. Sakar and E. Terry (2002). Multiperiod hedging with futures contracts. *Journal of Futures Markets*, 22, 1179–1203.

Lowengrub, P. and M. Melvin (2002). Before and after international cross-listing: An intraday examination of volume and volatility. *Journal of International Financial Markets, Institutions and Money*, 12, 139–155.

Lynch, P. and G. O. Zumbach (2003). Market heterogeneities and the causal structure of volatility. *Quantitative Finance*, 3, 320–331.

Mackinlay, A. C. (1995). Multifactor models do not explain deviations from the CAPM. *Journal of Financial Economics*, 38, 3–28.

Madsen, J. B. (2005). The Fisher hypothesis and the interaction between share returns, inflation and supply shocks. *Journal of International Money and Finance*, 24, 103–120.

Mallat, S. G. and W. L. Hwang (1992). Singularity detection and processing with wavelets. *IEEE Transportation Information Theory*, 38, 617–643.

Mallat, S. (1989). A theory for multiresolution signal decomposition: The wavelet representation. *IEEE Transactions on Pattern Analysis and Machine Intelligence*, 11, 674–693.

Mallat, S. (1999). *A Wavelet Tour of Signal Processing*. San Diago: Academic Press.

Mandelbrot, B. (1972). Statistical methodology for nonperiodic cycles: From the covariance to R/S analysis. *Annals of Economics and Social Measurement*, 1, 259–290.

Manimaran, P., P. K. Panigrahi and J. C. Parikh (2005). Wavelet analysis and scaling properties of time series. *Physics Review E*, 72, 046120.

Manimaran, P., P. K. Panigrahi and J. C. Parikh (2006). On estimation of Hurst scaling exponent and fractal behavior through discrete wavelets. Available at http://arxiv.org/abs/physics/060400.

McCarthy, J., R. DiSario, H. Saraoglu and H. Li (2004). Tests of long-range dependence in interest rates using wavelets. *Quarterly Journal of Economics and Finance*, 44, 180–189.

McCoy, E. J. and A. T. Walden (1996). Wavelet analysis and synthesis of stationary long-memory process. *Journal of Computational and Graphical Statistics*, 5, 1–31.

McGuinness, P. (1999). Volume effects in dual traded stocks: Hong Kong and London evidence. *Applied Financial Economics*, 9, 615–625.

McLeod, A. I. and K. W. Hipel (1978). Preservation of the rescaled adjusted range, 1: A reassessment of the Hurst phenomenon. *Water Resources Research*, 14, 491–508.

Newey, W. K. and K. D. West (1987). A simple, positive definite, heteroskedasticity and autocorrelation consistent covariance matrix. *Econometrica*, 55, 703–708.

Ogden, R. T. (1997). *Essential Wavelets for Statistical Applications and Data Analysis*. Boston: Birkhauser.

Ojanen, H. (1998). WAVEKIT: A wavelet toolbox for Matlab. Department of Mathematics, Rutgers University.

Pan, Z. and X. Wang (1998). A stochastic nonlinear regression estimator using wavelets. *Computational Economics*, 11, 89–102.

Pedersen, C. S. and T. Rudholm-Alfvin (2003). Selecting a risk-adjusted shareholder performance measure. *Journal of Asset Management*, 4, 152–172.

Pedersen, T. M. (1999). Understanding business cycles. Manuscript, Institute of Economics, University of Copenhagen.

Percival, D. B. (1995). On estimation of the wavelet variance. *Biometrika*, 82, 619–631.

Percival, D. B. and A. T. Walden (2000). *Wavelet Methods for Time Series Analysis*. Cambridge, UK: Cambridge University Press.

Pfingsten, A., P. Wagner and C. Wolferink (2004). An empirical investigation of rank correlation between different risk measures. *Journal of Risk*, 6, 55–74.

Pivetta, F. and R. Reis (2007). The persistence of inflation in the United States. *Journal of Economic Dynamics and Control*, 31, 1326–1358.

Politis, D. N. and H. White (2004). Automatic block-length selection for the dependent bootstrap. *Econometric Reviews*, 23, 53–70.

Politis, D. N. and J. P. Romano (1994). The stationary bootstrap. *Journal of the American Statistical Association*, 89, 1303–1313.

Priestley, M. B. (1992). *Spectral Analysis and Time Series*. San Diego: Academic Press.

Ramsey, J. B. (1999). The contribution of wavelets to the analysis of economic and financial data. Unpublished Paper, New York University.

Ramsey, J. B. (2002). Wavelets in economics and finance: Past and future. *Studies in Nonlinear Dynamics and Econometrics*, 6, 1–27.

Ramsey, J. B. and C. Lampart (1998a). Decomposition of economic relationships by timescale using wavelets. *Macroeconomic Dynamics*, 2, 49–71.

Ramsey, J. B. and C. Lampart (1998b). The decomposition of economic relationships by time scale using wavelets: Expenditure and income. *Studies in Nonlinear Dynamics and Economics*, 3, 23–42.

Ramsey, J. B. and D. J. Thomson (1999). A reanalysis of the spectral properties of some economic and financial time series. In Rothman, P. (ed.), *Nonlinear Time Series Analysis of Economic and Financial Data*. Boston, MA: Kluwer Academic Publishers, pp. 45–85.

Ramsey, J. B. and Z. Zhang (1997). The analysis of foreign exchange data using waveform dictionaries. *Journal of Empirical Finance*, 4, 341–372.

Ramsey, J. B., D. Usikov and D. Zaslavsky (1995). An analysis of US stock price behavior using wavelets. *Fractals*, 3, 377–389.

Robinson, P. M. (1995). Log-periodogram regression of time series with long range dependence. *Annals of Statistics*, 23, 1040–1072.

Rowe, W. W. and W. N. Davidson III (2000). Fund manager succession in closed-end mutual funds. *The Financial Review*, 35, 53–78.

Sanfilippo, G. (2003). Stocks, bonds and the investment horizon: A test of time diversification on the French market. *Quantitative Finance*, 3, 345–351.

Sargent, T. J. and C. A. Sims (1977). Business cycle modelling without pretending to have too much a priori economic theory. In Sims, C. A. (ed.), *New Methods in Business Cycle Research*. Federal Reserve Banks, Minneapolis.

Sarlan, H. (2001). Cyclical aspects of business cycle turning points. *International Journal of Forecasting*, 17, 369–382.

Schleicher, C. (2002). An introduction to wavelets for economists. Bank of Canada Working Paper, No. 2002–3.

Schotman, P. C. and M. Schweitzer (2000). Horizon sensitivity of the inflation hedge of stocks. *Journal of Empirical Finance*, 7, 301–305.

Serroukh, A. and A. T. Walden (2000a). Wavelet scale analysis of bivariate time series I: Motivation and estimation. *Journal of Nonparametric Statistics*, 13, 1–36.

Serroukh, A. and A. T. Walden (2000b). Wavelet scale analysis of bivariate time series II: Statistical properties for linear processes. *Journal of Nonparametric Statistics*, 13, 37–56.

Sharpe, W. F. (1994). The Sharpe ratio. *Journal of Portfolio Management*, (Fall), 49–58.

Siegel, A. F. and A. Woodgate (2007). Performance of portfolios optimized with estimation error. *Management Science*, 53, 1005–1015.

Siegel, J. J. (1998). *Stocks for the Long Run*. New York: McGraw-Hill.

Sims, C. A. (2001). Fiscal consequences for Mexico of adopting the dolloar. *Journal of Money, Credit and Banking*, 33, 597–616.

Smith, K. L. (1999). Major world equity market integration a decade after the 1987 crash: Evidence from cross spectral analysis. *Journal of Business Finance and Accounting*, 26, 365–392.

Smith, K. L. (2001). Pre- and post-1987 crash frequency domain analysis among Pacific Rim equity markets. *Journal of Multinational Financial Management*, 11, 69–87.

Solnik, B. and V. Solnik (1997). A multi-country test of the Fisher model for stock returns. *Journal of International Financial Markets, Institutions, and Money*, 7, 289–301.

Stock, J. and M. W. Watson (1999). Forecasting inflation. *Journal of Monetary Economics*, 44, 293–335.

Strang, G. and T. Nguyen (1996). *Wavelets and Filter Banks*. Wellesley, MA: Wellesley–Cambridge Press.

Sun, T. S., S. Sundaresan and C. Wang (1993). Interest rate swaps — An empirical investigation. *Journal of Financial Economics*, 34, 77–99.

Tkacz, G. (2001). Estimating the fractional order of integration of interest rates using a wavelet OLS estimator. *Studies in Nonlinear Dynamics and Econometrics*, 5(1), 1–14.

Trichet, J.-C. (2001). The euro after two years. *Journal of Common Market Studies*, 39, 1–13.

Valkonov, R. (2003). Long-horizon regressions: Theoretical results and applications. *Journal of Financial Economics*, 68, 201–232.

Vassalou, M. (2003). News related to future GDP growth as a risk factor in equity returns. *Journal of Financial Economics*, 68, 47–73.

Venetis, I. A. and D. Peel (2005). Non-linearity in stock index returns: The volatility and serial correlation relationship. *Economic Modelling*, 22, 1–19.

Wang, C. and S. S. Low (2003). Hedging with foreign currency denominated stock index futures: Evidence from the MSCI Taiwan index futures market. *Journal of Multinational Financial Management*, 13, 1–17.

Wang, S. S., O. M. Rui and M. Firth (2002). Return and volatility of dually-traded stocks: The case of Hong Kong. *Journal of International Money and Finance*, 21, 265–293.

Werner, I. M. and A. W. Kleidon (1996). UK and US trading British cross-listed stocks: An intraday analysis of market integration. *Review of Financial Studies*, 2, 619–664.

Whitcher, B. and M. J. Jensen (2000). Wavelet estimation of a local long memory parameter. *Exploration Geophysics*, 31, 89–98.

Whitcher, B., P. Guttorp and D. B. Percival (2000). Wavelet analysis of covariance with application to atmospheric time series. *Journal of Geophysical Research*, 105, 14,941–14,962.

Index